Roger Lee and Naohiro Ishii (Eds.)

Software Engineering, Artificial Intelligence, Networking and Parallel/Distributed
Computing

Studies in Computational Intelligence, Volume 209

Editor-in-Chief

Prof. Janusz Kacprzyk
Systems Research Institute
Polish Academy of Sciences
ul. Newelska 6
01-447 Warsaw
Poland
E-mail: kacprzyk@ibspan.waw.pl

Further volumes of this series can be found on our homepage: springer.com

Vol. 188. Agustín Gutiérrez and Santiago Marco (Eds.)
Biologically Inspired Signal Processing for Chemical Sensing,
2009
ISBN 978-3-642-00175-8

Vol. 189. Sally McClean, Peter Millard, Elia El-Darzi and Chris Nugent (Eds.)
Intelligent Patient Management, 2009
ISBN 978-3-642-00178-9

Vol. 190. K.R. Venugopal, K.G. Srinivasa and L.M. Patnaik
Soft Computing for Data Mining Applications, 2009
ISBN 978-3-642-00192-5

Vol. 191. Zong Woo Geem (Ed.)
Music-Inspired Harmony Search Algorithm, 2009
ISBN 978-3-642-00184-0

Vol. 192. Agus Budiyono, Bambang Riyanto and Endra Joelianto (Eds.)
Intelligent Unmanned Systems: Theory and Applications, 2009
ISBN 978-3-642-00263-2

Vol. 193. Raymond Chiong (Ed.)
Nature-Inspired Algorithms for Optimisation, 2009
ISBN 978-3-642-00266-3

Vol. 194. Ian Dempsey, Michael O'Neill and Anthony Brabazon (Eds.)
Foundations in Grammatical Evolution for Dynamic Environments, 2009
ISBN 978-3-642-00313-4

Vol. 195. Vivek Bannore and Leszek Swierkowski
Iterative-Interpolation Super-Resolution Image Reconstruction:
A Computationally Efficient Technique, 2009
ISBN 978-3-642-00384-4

Vol. 196. Valentina Emilia Balas, János Fodor and Annamária R. Várkonyi-Kóczy (Eds.)
Soft Computing Based Modeling in Intelligent Systems, 2009
ISBN 978-3-642-00447-6

Vol. 197. Mauro Birattari
Tuning Metaheuristics, 2009
ISBN 978-3-642-00482-7

Vol. 198. Efrén Mezura-Montes (Ed.)
Constraint-Handling in Evolutionary Optimization, 2009
ISBN 978-3-642-00618-0

Vol. 199. Kazumi Nakamatsu, Gloria Phillips-Wren, Lakhmi C. Jain, and Robert J. Howlett (Eds.)
New Advances in Intelligent Decision Technologies, 2009
ISBN 978-3-642-00908-2

Vol. 200. Dimitri Plemenos and Georgios Miaoulis *Visual Complexity and Intelligent Computer Graphics Techniques Enhancements,* 2009
ISBN 978-3-642-01258-7

Vol. 201. Aboul-Ella Hassanien, Ajith Abraham, Athanasios V. Vasilakos, and Witold Pedrycz (Eds.)
Foundations of Computational Intelligence Volume 1, 2009
ISBN 978-3-642-01081-1

Vol. 202. Aboul-Ella Hassanien, Ajith Abraham, and Francisco Herrera (Eds.)
Foundations of Computational Intelligence Volume 2, 2009
ISBN 978-3-642-01532-8

Vol. 203. Ajith Abraham, Aboul-Ella Hassanien, Patrick Siarry, and Andries Engelbrecht (Eds.)
Foundations of Computational Intelligence Volume 3, 2009
ISBN 978-3-642-01084-2

Vol. 204. Ajith Abraham, Aboul-Ella Hassanien, and André Ponce de Leon F. de Carvalho (Eds.)
Foundations of Computational Intelligence Volume 4, 2009
ISBN 978-3-642-01087-3

Vol. 205. Ajith Abraham, Aboul-Ella Hassanien, and Václav Snášel (Eds.)
Foundations of Computational Intelligence Volume 5, 2009
ISBN 978-3-642-01535-9

Vol. 206. Ajith Abraham, Aboul-Ella Hassanien, André Ponce de Leon F. de Carvalho, and Václav Snášel (Eds.)
Foundations of Computational Intelligence Volume 6, 2009
ISBN 978-3-642-01090-3

Vol. 207. Santo Fortunato, Giuseppe Mangioni, Ronaldo Menezes, and Vincenzo Nicosia (Eds.)
Complex Networks, 2009
ISBN 978-3-642-01205-1

Vol. 208. Roger Lee, Gongzu Hu, and Huaikou Miao (Eds.)
Computer and Information Science 2009, 2009
ISBN 978-3-642-01208-2

Vol. 209. Roger Lee and Naohiro Ishii (Eds.)
Software Engineering, Artificial Intelligence, Networking and Parallel/Distributed Computing, 2009
ISBN 978-3-642-01202-0

Roger Lee and Naohiro Ishii (Eds.)

Software Engineering, Artificial Intelligence, Networking and Parallel/Distributed Computing

Springer

Prof. Roger Lee
Computer Science Department
Central Michigan University
Pearce Hall 413
Mt. Pleasant, MI 48859
USA
Email: lee1ry@cmich.edu

Naohiro Ishii
Department of Information Science
Aichi Institute of Technology
Toyota, Japan
Email: ishii@in.aitech.ac.jp

ISBN 978-3-642-10173-1   e-ISBN   978-3-642-01203-7

DOI 10.1007/978-3-642-01203-7

Studies in Computational Intelligence   ISSN 1860949X

Typeset & Cover Design: Scientific Publishing Services Pvt. Ltd., Chennai, India.

Printed in acid-free paper

9 8 7 6 5 4 3 2 1

springer.com

Preface

The purpose of the 10th ACIS International Conference on Software Engineering Artificial Intelligence, Networking and Parallel/Distributed Computing (SNPD 2009), held in Daegu, Korea on May 27–29, 2009, the 3rd International Workshop on e-Activity (IWEA 2009) and the 1st International Workshop on Enterprise Architecture Challenges and Responses (WEACR 2009) is to aim at bringing together researchers and scientist, businessmen and entrepreneurs, teachers and students to discuss the numerous fields of computer science, and to share ideas and information in a meaningful way. Our conference officers selected the best 24 papers from those papers accepted for presentation at the conference in order to publish them in this volume. The papers were chosen based on review scores submitted by members of the program committee, and underwent further rounds of rigorous review.

In chapter 1, Igor Crk and Chris Gniady propose a network-aware energy management mechanism that provides a low-cost solution that can significantly reduce energy consumption in the entire system while maintaining responsiveness of local interactive workloads. Their dynamic mechanisms reduce the decision delay before the disk is spun-up, reduce the number of erroneous spin-ups in local workstations, decrease the network bandwidth, and reduce the energy consumption of individual drives.

In chapter 2, Yoshihito Saito and Tokuro Matsuo describe a task allocation mechanism and its performance concerning with software developing. They run simulations and discuss the results in terms of effective strategies of task allocation.

In chapter 3, Wei Liao et al. present a novel directional graph model for road networks to simultaneously support these two kinds of continuous k-NN queries by introducing unidirectional network distance and bidirectional network distance metrics. Experimental results are contrasted with the results existing algorithms including IMA and MKNN algorithms.

In chapter 4, Liya Fan et al. present a self-adaptive greedy scheduling scheme to solve a Multi-Objective Optimization on Identical Parallel Machines. The primary objective is to minimize the makespan, while the secondary objective is to make the schedule more stable. To demonstrate their claim they provide results derived from their experimentation.

In chapter 5, Tomohiko Takagi et al. show that a UD (Usage Distribution) coverage criterion based on an operational profile measures the progress of software testing from users' viewpoint effectively. The goal of the work is to describe the definitions and examples of UD coverage.

In chapter 6, Vahid Jalali and Mohammad Reza Matash Borujerdi propose a method of semantic information retrieval that can be more concise based on adding specially selected extra criterion. Using the results from their studies they compare the effectiveness against that of keyword based retrieval techniques.

In chapter 7, Tomas Bures et al. perform a study on the transparency of connectors in a distributed environment. They evaluate the feasibility of making the remote communication completely transparent and point out issues that prevent the transparency and analyze the impact on components together with possible tradeoffs.

In chapter 8, Dong-sheng Liu and Wei Chen propose an integrated model to investigate the determinants of user mobile commerce acceptance on the basis of innovation diffusion theory, technology acceptance model and theory of planned behavior considering the important role of personal innovativeness in the initial stage of technology adoption. The proposed model was empirically tested using data collected from a survey of MC consumers.

In chapter 9, Wenying Feng et al. demonstrate a cache prefetching method that builds a Markov tree from a training data set that contains Web page access patterns of users, and make predictions for new page requests by searching the Markov tree. Using simulations based on Web access logs they are able to demonstrate the effectiveness of this methodology.

In chapter 10, Haeng-Kon Kim and Roger Y. Lee design and implement a web-using mining system using page scrolling to track the position of the scroll bar and movements of the window cursor regularly within a window browser for real-time transfer to a mining server and to analyze user's interest by using information received from the analysis of the visual perception area of the web page.

In chapter 11, WU Kui et al. propose a new method for similarity computation based on Bayesian Estimation. They implement analyze the performance using WordNet.

In chapter 12, Ghada Sharaf El Din et al. use a predictive modeling approach to develop models using political factors, social-economic factors, degree of development, etc. to predict an index derived from the institutional and normative aspects of democracy.

In chapter 13, Mohsen Jafari Asbagh, Mohsen Sayyadi, Hassan Abolhassani propose a step in processing blogs for blog mining. They use a shallow summarization method for blogs as a preprocessing step for blog mining which benefits from specific characteristics of the blogs including blog themes, time interval between posts, and body-title composition of the posts. We use our method for summarizing a collection of Persian blogs from PersianBlog hosting and investigate its influence on blog clustering.

In chapter 14, Chia-Chu Chiang and Roger Lee propose an approach using formal concept analysis to extract objects in a non-object oriented programs. Their research is a collaborative work in which they are trying to draw the benefit by using formal concept analysis and apply the analysis results to the field of software maintenance.

In chapter 15, Tokuro Matsuo and Norihiko Hatano employ qualitative methods in place of quantitative methods for flood analysis. Their qualitative simulation model is constructed as a causal graph model. This system is an attempt to better suit this flood analysis for naïve users and novices.

In chapter 16, Oliver Strong et al. propose a simple technique of UML markup that would help bridge the gaps of UML illiteracy and diagram ambiguity. This is done by using a simple textual markup inscribed at the XML level; they are able to demonstrate with a simple test case the ease and verbosity of transparent markup applied to UML diagrams.

In chapter 17, Jae-won Lee et al. propose a new methodology for solving the item sparsity problem by mapping users and items to a domain ontology. Their method uses a semantic match with the domain ontology, while the traditional collaborative filtering uses an exact match to find similar users. They perform several experiments to show the benefits of their methods over those of exact matching.

In chapter 18, Haeng-Kon Kim works towards creating a system that allows purchasers to discriminate between software based on its tested suitability. To do this he builds a system that purchasers can effectively select a software package suitable for their needs, using quality test and certification process for package software and developing a Test Metric and application method.

In chapter 19, Olga Ormandjieva et al. formalize the EI modeling and performance control in a single formal framework based on representational theory of measurement and category theory.

In chapter 20, Myong Hee Kim and Man-Gon Park propose Bayesian statistical models for observed active and sleep times data of sensor nodes under the selected energy efficient CSMA contention-based MAC protocols in consideration of the system effectiveness in energy saving in a wireless sensor network. Accordingly, we propose Bayes estimators for the system energy saving effectiveness of the wireless sensor networks by use of the Bayesian method under the conjugate prior information.

In chapter 21, Chang-Sun Shin et al. propose a system for the management of environmental factors in livestock and agriculture industries. They provide a three layer system that makes use of natural energy sources to monitor environmental factors and provide a reporting functionality.

In chapter 22, Jeong-Hwan Hwang et al. expand the capabilities of Ubiquitous-Greenhouse by creating a mesh network which is used to overcome the problem of sensor node energy utilization. They measure the successfulness of this approach in terms of energy consumption and data transmission efficiency.

In chapter 23, Byeongdo Kang and Roger Y. Lee present an architecture style for switching software. They attempt to make switching software more maintainable by suggesting a hierarchical structure based on the characteristics of switching software.

In chapter 24, the final chapter, Roman Neruda studies the problem of automatic configuration of agent collections. A solution combining logical resolution system and evolutionary algorithm is proposed and demonstrated on a simple example.

It is our sincere hope that this volume provides stimulation and inspiration, and that it will be used as a foundation for works yet to come.

May 2009   Roger Lee
                                                                    Naohiro Ishii

Contents

List of Contributors

Mohsen Jafari Asbagh
Sharif University, Iran
m_jafari@ce.sharif.edu

Mohammad Reza Matash Borujerdi
Amirkabir University of Technology,
Iran
borujerm@aut.ac.ir

Tomas Bures
Charles University, Czech Republic
bures@dsrg.mff.cuni.cz

Wei Chen
Zhejiang Gongshang University, China
true1.chen@gmail.com

Chia-Chu Chiang
University of Arkansas at Little Rock,
USA
cxchiang@ualr.edu

Igor Crk
University of Arizona, USA
icrk@cs.arizona.edu

Ghada Sharaf El Din
Central Michigan University, USA

Liya Fan
Chinese Academy of Sciences, China
fanliya@ict.ac.cn

Moataz Fattah
Central Michigan University, USA

Wenying Feng
Trent University
wfeng@trentu.ca

Zengo Furukawa
Kagawa University, Japan
zengo@eng.kagawa-u.ac.jp

Chris Gniady
University of Arizona, USA
gniady@cs.arizona.edu

Josef Hala
Charles University, Czech Republic
hala@dsrg.mff.cuni.cz

Norihiko Hatano
Yamagata University, Japan

Petr Hnetynka
Charles University, Czech Republic
hnetynka@dsrg.mff.cuni.cz

Gongzhu Hu
Central Michigan University, USA
hu1g@cmich.edu

Jeong-Hwan Hwang
Sunchon National University, Korea
jhwang@mail.sunchon.ac.kr

Hyuk-Jin Im
Sunchon National University, Korea
polyhj@mail.sunchon.ac.kr

Vahid Jalali
Amirkabir University of
Technology, Iran
vjalali@aut.ac.ir

Wang Jianyu
Nanjing University of Science &
Technology, China

Su-Chong Joo
Wonkwang University, Korea
scjoo@wonkwang.ac.kr

Byeongdo Kang
Daegu University, Korea
bdkang@daegu.ac.kr

Hyun-Joong Kang
Sunchon National University, Korea
hjkang@mail.sunchon.ac.kr

Haeng-Kon Kim
Catholic Univ. of Daegu, Korea
hangkon@cu.ac.kr

Myong Hee Kim
Pukyong National University, Korea
mhgold@naver.com

Stan Klasa
University, Montreal, Canada
klasa@cse.concordia.ca

Wu Kui
Nanjing University of Science &
Technology, China

Carl Lee
Central Michigan University, USA
Carl.lee@cmich.edu

Jae-won Lee
Seoul National University, Korea
lyonking@europa.snu.ac.kr

Roger Y. Lee
Central Michigan University, USA
lee1ry@cmich.edu

Sang-goo Lee
Seoul National University, Korea
sglee@europa.snu.ac.kr

Yong-Woong Lee
Sunchon National University, Korea
ywlee@sunchon.ac.kr

Wei Liao
Naval University of Engineering, China
liaoweinudt@yahoo.com.cn

Guo Ling
Nanjing University of Science &
Technology, China

Dong-sheng Liu
Zhejiang Gongshang University, China
lds1118@mail.zjgsu.edu.cn

Zhiyong Liu
Chinese Academy of Sciences, China
zyliu@ict.ac.cn

Shushuang Man
Southwest Minnesota State
University, USA
mans@southwestmsu.edu

Tokuro Matsuo
Yamagata University, Japan
matsuo@tokuro.net

Victoria Mikhnovsky
University, Montreal, Canada
tori.mikhnovsky@hotmail.com

Takashi Mitsuhashi
JustSystems Corporation, Japan

Masayuki Muragishi
JustSystems Corporation, Japan

Kwang-Hyun Nam
Seoul National University, Korea
nature1226@europa.snu.ac.kr

Roman Neruda
Academy of Sciences of the Czech
Republic
roman@cs.cas.cz

Kazuya Nishimachi
JustSystems Corporation, Japan

Olga Ormandjieva
University, Montreal, Canada
ormandj@cse.concordia.ca

Man-Gon Park
Pukyong National University, Korea
mpark@pknu.ac.kr

Yoshihito Saito
Yamagata University, Japan
saito2007@e-activity.org

Mohsen Sayyadi
Sharif University, Iran
m_sayyadi@ce.sharif.edu

Chang-Sun Shin
Sunchon National University, Korea
csshin@sunchon.ac.kr

Choun-Bo Sim
Sunchon National University, Korea
cbsim@sunchon.ac.kr

Oliver Strong
Central Michigan University, USA
stron1om@cmich.edu

Tomohiko Takagi
Kagawa University, Japan
takagi@ismail.eng.
kagawa-u.ac.jp

Gongming Wang
Chinese Academy of Sciences, China
wanggongming@ict.ac.cn

Xiaoping Wu
National University of Defense
Technology, China

Zhuo Xianzhong
Nanjing University of Science &
Technology, China

Chenghua Yan
National University of Defense
Technology, China

Hyun Yoe
Sunchon National University, Korea
yhyun@sunchon.ac.kr

Bo Yuan
Shanghai Jiao Tong University, China
yuanbo@cs.sjtu.edu.cn

Fa Zhang
Chinese Academy of Sciences, China
zf@ncic.ac.cn

Zhinong Zhong
National University of Defense
Technology, China

Network-Aware Program-Counter-Based Disk Energy Management

Igor Crk and Chris Gniady

Abstract. Reducing energy consumption is critical to prolonging the battery life in portable devices. With rising energy costs and increases in energy consumption by devices in stationary systems energy expenses of large corporations can annually reach into millions of dollars. Subsequently, energy management has also become important for desktop machines in large scale organizations. While energy management techniques for portable devices can be utilized in stationary systems, they do not consider network resources readily available to stationary workstations. We propose a network-aware energy management mechanism that provides a low-cost solution that can significantly reduce energy consumption in the entire system while maintaining responsiveness of local interactive workloads. The key component of the system is a novel program-context-based bandwidth predictor that accurately predicts application's bandwidth demand for file server/client interaction. Our dynamic mechanisms reduce the decision delay before the disk is spun-up, reduce the number of erroneous spin-ups in local workstations, decrease the network bandwidth, and reduce the energy consumption of individual drives.

1 Introduction

Dynamic energy reduction techniques have been extensively studied with the aim of extending battery life and prolonging the operation of portable devices. Considering that end-systems in large-scale enterprise environments are already dependent, to some degree, on centralized storage, we see an opportunity for reducing energy

Igor Crk
University of Arizona, Tucson AZ 85721
e-mail: icrk@cs.arizona.edu

Chris Gniady
University of Arizona, Tucson AZ 85721
e-mail: gniady@cs.arizona.edu

R. Lee, N. Ishii (Eds.): Software Engineering, Artificial Intelligence, SCI 209, pp. 1–13.
springerlink.com   © Springer-Verlag Berlin Heidelberg 2009

Fig. 1 Anatomy of an idle
period

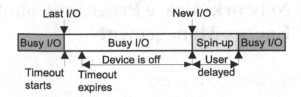

consumption in both portable and desktop systems that rely on this resource. We
propose a bandwidth demand predictor and design an end-system energy management mechanism that can be successfully used within large enterprise networks.
The techniques we develop here are applicable to many system components, but our
focus is on the hard drive, since it is responsible for a significant portion of overall
system energy. The critical issues are keeping the hard drive spinning only when
necessary, relying on the network for common data access, while at the same time
avoiding network congestion.

We propose the Program Context Bandwidth Predictor (PCBP) for dynamic I/O
bandwidth prediction that meets this need. The PCBP idea is supported by recent
research regarding I/O context correlation [8, 9]. In this paper, we argue that there is
a strong correlation between the application context and the I/O bandwidth demand
generated in by the context. Furthermore, we exploit this correlation to accurately
predict bandwidth demands from the application and design a network-aware power
management system which dynamically adjusts the level of reliance on network
storage based on bandwidth availability and predicted future bandwidth demand.
The proposed mechanism is compared to the previously proposed prediction mechanism in Self-Tuning Power Management (STPM) [14]. Our results show that PCBP
mechanisms achieve higher accuracy, better responsiveness, and more efficient network bandwidth utilization while improving energy efficiency.

2 Background

Manufacturers recommend spinning down the power hungry hard disks after some
period of idleness [6, 11]. Fig. 1 shows the anatomy of an idle period. After the last
request, a timer is started and the device is shut down once the timer expires unless
there is additional disk activity occuring during the timeout period. The disk remains powered down until a new request arrives. However, accelerating the platters
requires more energy than keeping them spinning while the disk is idle. Therefore,
the time during which the device is off has to be long enough to offset the extra energy needed for the shutdown and spin-up sequence. This time is commonly referred
to as the *breakeven time*, and is usually on the order of a few seconds. Eliminating
shutdowns that not only waste energy but also significantly delay user requests is
critical to conserving energy and reducing interactive delays.

Timeout-based mechanisms are simple to implement but waste energy while waiting for a timeout to expire. Consequently, dynamic predictors that shut down the device much earlier than the timeout mechanisms have been proposed to address the energy consumption of the timeout period [5, 12, 18]. Stochastic modeling techniques have also been applied to modeling the idle periods in applications and shut down the disk based on the resulting models [3, 4, 15, 17]. Additional heavy weight approaches have relied on application developers to insert hints about upcoming activity [7, 10, 13, 19] or have suggested compile-time optimizations for I/O reshaping and disk access synchronization.

To reduce the burden of hint insertion on the programmer, automatic generation of application hints was proposed in Program Counter Access Predictor (PCAP) [9]. PCAP exploits the observation that I/O activity is caused by unique call sites within the applications. Depending on user interactions with the application, I/O sequences are generated by different call sites. Therefore, the activity described by a call site can be correlated to an idle period that follows a sequence of I/O operations. The program context of the idle period provides more information to the predictor, resulting in high accuracy and good energy savings.

In addressing both spin-up delays and energy efficiency, BlueFS [14] proposes that data should be fetched from alternate sources to mask the disk latency. The recent use of flash memory in hybrid disk drives by Samsung [16] provides similar energy savings. BlueFS uses ghost hints [1, 2] to predict when the disk should be spun-up. Ghost hints are issued to the disk for each item fetched from the network, and after receiving a few such hints in a short time, the disk is spun-up and takes over serving requests. A ghost hint discloses the potential time and energy that would have been saved if the disk was in the ideal power mode, i.e. the platters are spinning and the disk is ready to serve data. A running total of benefits is maintained and increased every time a hint is received. The total is also decremented by the energy cost of remaining in idle mode since the last ghost hint was issued. The disk is spun-up when the total exceeds a threshold. Throughout the spin-up period, data is fetched from the network, making disk power management transparent to the user. Furthermore, requests with low I/O activity are fetched entirely from the server, avoiding disk spin-ups altogether. By eliminating disk spin-ups we can significantly reduce both the client's power consumption and the wear-and-tear of the disk's mechanical components.

However, BlueFS's ghost hinting mechanisms suffer from drawbacks similar to those of the timeout-based shutdown mechanisms. First, the delay of spinning up the disk can result in longer delays for high bandwidth I/O requests. Second, some requests that result in a disk spin-up according to ghost hinting mechanisms may be satisfied before the disk spins up, resulting in no disk activity following the spin-up. Therefore, a dynamic predictor that can exploit a large history of I/O events and provide immediate spin-up predictions has the potential to improve the accuracy and timeliness of predictions.

Authors of PCAP have shown that using call sites to describe the behavior of different parts of the application can lead to a very accurate shutdown predictor. Different parts of the application exhibit different I/O behavior, therefore uniquely correlating application parts to the resulting I/O behavior should improve spin-up predictions, as well. This paper shows that we can accurately correlate the I/O bandwidth demand to the call sites to improve disk spin-up predictions.

3 Design

The key idea behind PCBP is that *there is a strong correlation between the program context that initiated the sequence of I/O operations and the resulting I/O activity.* We exploit this correlation to control the disk activity of desktop systems in a large enterprise environment, under the assumption that network storage is available and a relatively small portion of available bandwidth can be apportioned to support I/O for local applications.

3.1 Program Context Correlation

We define a busy period as the I/O activity between two idle periods that are longer than the breakeven time. To uniquely describe the program context, PCBP calculates the signature of the call site that initiated the I/O by traversing the application call stack and recording the program counter of each function call on the stack [9]. PCBP collects and records a single signature for each busy period, minimizing the overhead of the system call execution. Once the signature is recorded, PCBP begins collecting statistics about the current busy period.

PCBP stores the number of bytes requested (*busy-bytes*) during the busy period and the length of the busy period (*busy-time*) with each signature. PCBP updates and maintains the *busy-bytes* and *busy-time* variables per application, as shown in Fig. 2. Each time a new I/O request arrives, PCBP calculates the length of the idle period between the completion of the last I/O request and the beginning of the current one. If the idle period is shorter than the breakeven time, it is considered part of the busy

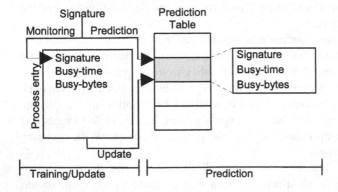

Fig. 2 Prediction and training structures

period. An idle period longer than the breakeven time indicates the beginning of a new busy period.

The prediction table is organized as a hash table indexed by the call site signature. The table is shared among processes to allow table reuse across multiple executions of the same application as well as concurrent executions of the same application. Table reuse reduces training for future invocations of an application, and can be easily retained in the kernel across multiple executions of the application due to its small size. Table reuse was successfully exploited in previous program context based predictors [9, 8]. Additionally, the table can be stored locally or remotely in order to eliminate training following a reboot.

3.2 Disk Spin-Up Prediction

PCBP predicts I/O activity that will follow a call site preceding the first I/O of a busy period. Continuing to monitor I/O activity during the busy period ensures proper handling of mispredictions. PCBP's predictions of upcoming I/O traffic and busy period duration are used to determine whether the disk should be spun up or the file server should satisfy the request instead. The file server is assumed to contain all user files and application files and sufficient memory to cache them.

To spin up, PCBP first considers the possibility of bursty traffic. In this case, if a request can be satisfied from the file server in the time it takes to spin up the disk in the local workstation, the disk remains down. In making the decision, PCBP considers the predicted length of the busy period. If the predicted length is shorter than the time it takes to spin up the disk, PCBP considers the amount of I/O traffic that is to be fetched during the busy period. If the I/O requests can be satisfied from the server, given a particular server load, in the time it would take to spin the disk up, the disk is not spun up and the requests are serviced by the file server. Otherwise, PCBP spins the disk up immediately. While the disk is spinning up, I/O requests are serviced by the file server. Once the disk is ready, it takes over and services the remaining requests.

The second step in spin-up prediction considers steady traffic with low bandwidth demand. In this case, a low bandwidth prediction in the primary PCBP predictor supresses the issuing of ghost hints, leaving the backup mechanism to make predictions only when no primary PCBP prediction is available. A spin-up occuring for low-bandwidth activity periods is considered a miss, since the requested data can be served by the server with little to no performance impact. A beneficial spin-up reduces the delay associated with serving data from the server when the available bandwidth is saturated. To prevent excessive reliance on network storage during training and mispredictions, PCBP uses the BlueFS ghost hinting mechanism as a backup predictor.

PCBP mechanisms use the same criteria for a spin-up decision in the case of both a single and multiple processes issuing I/O requests. A prediction for each process issuing I/O requests is sent to the Global Spin-up Coordinator (GSC), which considers the combination of all currently predicted I/O demands for each process in

Table 1 Applications and execution details

Applications	Number of Executions	Global Busy Periods	Local Busy Periods	Total Signatures
mozilla	30	361	997	222
mplayer	10	10	10	1
impress	29	180	632	110
writer	29	137	415	61
calc	26	150	591	67
xemacs	31	100	127	14

deciding whether to spin the disk up. As long as the sum of I/O activity arriving from processes can be satisfied by the given file server bandwidth, the disk remains in the off state. If GSC detects that the bandwidth demand will most likely drop below the available bandwidth in the time it would take to spin the disk up, the disk is not spun up. In addition, GSC monitors the traffic from all applications and compares it to the total predicted traffic. If there is no prediction available, such as during training, or if following a prediction to solely utilize the network for upcoming requests the amount of I/O observed exceeds the amount of I/O that was predicted, the backup predictor is activated.

PCBP relies on BlueFS mechanisms to synchronize the local file system to the file server. User generated data is transferred to the file server periodically, allowing for extended buffering in memory. In the case of machine failure, user data on the local machine can be restored to its correct state from the server. Further, the mirroring of user and application data from all workstations does not cause a major storage overhead, due to the relative homogeneity of enterprise computing environments. As shown by the BlueFS system study [14], the amount of user data that has to be synchronized with the server is quite small and should not result in excessive communication overheads or energy consumption.

4   Methodology

We evaluate the performance of PCBP and compare it to ghost hinting mechanisms in BlueFS using a trace-based simulator. Detailed traces of user-interactive sessions for each of a set of applications were obtained by a modified `strace` utility over a number of days. The modified `strace` utility allows us to obtain the PC, PID, access type, time, file descriptor, file name, and amount of data that is fetched for each I/O operation. The applications themselves are popular interactive productivity and entertainment applications familiar to Linux users.

Table 1 shows the details of the chosen applications' traces. As previously discussed, global busy periods are composed of one or more local busy periods. Per-process predictions are performed using the PC signatures at the beginning of each local busy period. GSC uses these predictions to generate a prediction for each global busy period.

Table 2 Energy consumption specifications, **a** for the WD2500JD hard disk drive (left), and **b** the CISCO 350 wireless network card (right)

State	Power
R/W	10.6W
Seek	13.25W
Idle	10W
Standby	1.7W

Transition	Energy
Spin-up	148.5J
Shutdown	6.4J

Transition	Delay
Spin-up	9 sec.
Shutdown	4 sec.
Breakeven	19 sec.

State	Power
Low	.39W
High	1.41W

Transition	Energy
Ready	.510J
Shutdown	.530J

Transition	Delay
Ready	.40 sec.
Shutdown	.41 sec.

Energy consumption and savings are calculated based on application behavior and the amount of time spent in a particular state. Table 2 shows the energy consumption profiles of a Western Digital Caviar 250.0GB Internal Hard Drive Model WD2500JD. and the Cisco 350 Network Interface Card.

We implemented the ghost hinting mechanisms according to the description in [14, 2], placing equal weight on response time and energy efficiency. A simple timeout-based shutdown predictor set to breakeven time provides the common basis for comparison of accuracy of the two mechanisms and does not introduce another dimension for evaluation.

The evaluation assumes that the available network bandwidth to the server is constrained to one thousandth of the total bandwidth of a 1 Gb network, chosen experimentally as the point after which further constraints on bandwidth availability don't result in significant energy savings for the local disk. An overabundance of available network bandwidth would keep the disk mostly in the off state, so by constricting the available bandwidth we illustrate that energy savings are possible when the available bandwidth is low.

5 Evaluation

5.1 Prediction Accuracy

Figure 3 compares the ability of Ghost Hints (GH) and PCBP to correctly predict the I/O activity during a busy period. The GH mechanism consists of ghost hints alone, therefore Fig. 3 shows only hits and misses for GH. PCBP mechanisms use ghost hints as a backup predictor during training and mispredictions. We define a hit as a disk spin-up followed by bandwidth demand that is higher than what is available from the server. Spinning up the disk wastes energy when few or no I/O requests are served following spin-up, hence these events are defined as misses.

We find that on average PCBP mechanisms make 17% more correct disk spin-ups than GH, with an average of 76% fewer misses than GH. The primary predictor is responsible for a majority of the coverage, attaining on average 88% of the hits. To avoid a possibility of inheriting the large misprediction rate of GH, we use a 2-bit saturating counter as a confidence estimator to guide PCBP in using the backup predictor. The resulting misprediction rate of the backup predictor is on average 18%.

5.2 Server and Disk Utilization

Figure 4 shows the fraction of data served from the server for GH and PCBP. It is important to emphasize that the reliance on a network server for user data should not adversely affect the operation of the local system nor other systems accessing network resources. We observe that the amount of data fetched from the disk was greater when using PCBP mechanisms for all applications, excluding *Mplayer*. Conversely, we observe that using PCBP mechanisms, the load on the server is on average 3% less than with GH. We can conclude that using PCBP mechanisms, the server can either increase the number of machines it can support or provide a higher quality service to the same number of machines.

Figure 3 shows that PCBP mechanisms initiate fewer spin-ups while serving more data from the disk, as shown in Fig. 4. This implies that PCBP mechanisms are more efficient in serving data from the disk as shown in Fig. 5. Figure 5 shows the average amount of data served from the local drive with PCBP and GH per beneficial disk spin-up. On average, PCBP mechanisms result in serving 7% more data than GH following each disk spin-up. Higher efficiency in PCBP mechanisms results from the PCBP spinning up the disk sooner than GH, therefore serving more data from the disk in a particular busy period. In all applications, the disk spin-ups were consistently more beneficial when using PCBP mechanisms.

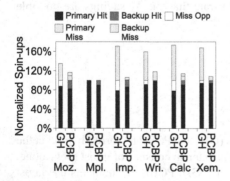
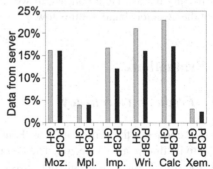

Fig. 3 Comparison spin-up predictor accuracies, normalized to the number of spin-ups

Fig. 4 Percent of data served by the server

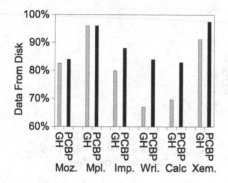

Applications	GH Ave. Delay	PCBP Ave. Delay
mozilla	10.46s	5.42s
mplayer	1.37s	0.01s
impress	6.17s	4.60s
writer	3.90s	2.26s
calc	5.48s	4.78s
xemacs	2.42s	1.15s

Fig. 5 Average amount of data fetched per beneficial disk spin-up

Fig. 6 Average delay in seconds before the disk has begun spinning up

5.3 Response Time

Table 6 details the average delay time before the decision to spin the disk up is made by GH and PCBP. PCBP's timely prediction and infrequent reliance on the backup predictor ensure that the PCBP spin-up delay time is on average 47% shorter than that of GH mechanisms.

5.4 Predictability of Process Activity

The distribution of I/O activity for all collected signatures is shown in Fig. 7 as the distance from the mean. A distance of one represents the range of values equivalent to one twentieth of the average expected I/O activity for all traced applications. Each increment is equivalent to 12kB. We observe that 40% of all activity equals the mean for a signature each time that signature is observed. Further, more than 80% of activity is less than 120kB from the mean, while the average local-busy activity for all traced processes is 240kB. Irregularities in the distribution are due to the nature of interactive applications, for example, one signature may be used to record the activity associated with accessing multiple user-specified files of significantly different sizes.

Figure 8 shows that the aggregate time elapsed between requests in a single busy period of a process shows small variability, as did Fig. 7 for the number of bytes fetched. Distances from the mean are computed at microsecond accuracy. In this case, we find that over 85% of measured aggregate times fall within 10ms of the mean for a particular signature. The predictability of bytes fetched during I/O activity and its elapsed time allows us to make accurate decisions about future activity and its bandwidth requirements.

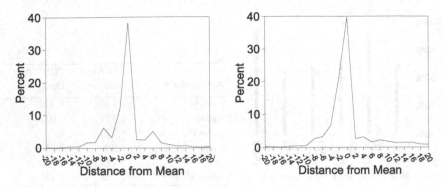

Fig. 7 Distribution of bytes fetched during **Fig. 8** Distribution of I/O activity by duration
I/O activity

5.5 Energy Savings

Figure 9 compares the energy-delay products of GH and PCBP, normalized to the
energy-delay of the on-demand spin-up mechanism. All predictors evaluated in this
figure shut down the disk after the breakeven time has passed since the last request has been served. The energy consumption is calculated from a combination of
power-cycle energy, consumed when the disk spins up or shuts down, idle energy,
consumed when the disk is spinning idly, and active energy, consumed when the
disk is actively seeking, reading or writing.

Overall, PCBP shows an average of 40% improvement in energy-delay product over the demand-based mechanism, due to its overall 52% improvement in energy consumption and timeliness of spin-ups in instances where bandwidth demand

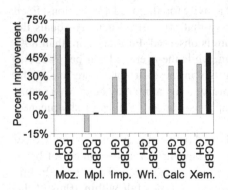

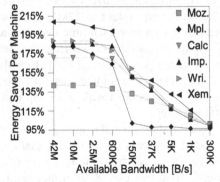

Fig. 9 Hard drive energy-delay product improvement normalized to demand-based disk
spin-up

Fig. 10 Percent of energy saved per network-
attached machine as available network
bandwidth decreases, normalized to the
demand-based predictor

exceeds the allowable network limit. GH performs reasonably well with an average 30% improvement in energy-delay product over the demand-based mechanism. This corresponds to the observations seen in Figs. 4 and 5.

Figure 10 presents energy savings for one network-attached machine given the constrained available server bandwidth. We use a standard workstation as a file server with a measured peak bandwidth of 42 MB/s. This is shown in Fig. 10 as the maximum bandwidth available in our experments. When the available bandwidth is systematically restricted, a drop in energy savings occurs at 150 KB/s for most applications, suggesting that at this point the I/O generated by the attached machine has begun to saturate the allocated bandwidth.

6 Conclusion

Energy consumption plays a significant role not only in portable devices but also in large enterprises due to rising energy costs and increases in energy consumption by stationary systems. In this paper, we proposed a novel program-context-bandwidth predictor and utilize the predictor in designing energy management mechanisms for large scale enterprise environments.

The paper presented evidence of a strong correlation between program context and an application's bandwidth demand. The proposed PCBP mechanisms accurately correlate bandwidth demand during busy periods with the program context that initiated the busy period, and thus have the potential to make much more accurate predictions than previous schemes. Compared to ghost hinting mechanisms that predict I/O behavior based only on most recent history, the PCBP mechanism offers several advantages: (1) it can accurately predict the I/O activity of the new busy period during the first request that initiated the busy period; (2) it spins up the disk immediately upon predicting the high activity busy period, eliminating the prediction delays due to recent history collection; (3) it significantly reduces number of unnecessary spin-ups reducing impact of energy management on disk reliability.

Our evaluations using desktop applications commonly encountered in enterprise environments show that compared to ghost hinting mechanisms in BlueFS, PCBP mechanisms reduce unnecessary disk spin-ups on average by 76%. The predicted spin-ups are more efficient by serving on average 6.8% more data per spin-up than ghost hints. As a result, PCBP mechanisms improve the energy-delay product on average by an additional 10% for an overall improvement of 40% from the popular timeout-based mechanism. With respect to energy consumption, PCBP is only 4% from a perfect spin-up predictor.

PCBP can be combined with PCAP [9] to fully integrate context based shutdown and spin-up prediction. Furthermore, it is possible to use PCBP to predict shutdown due to accurate prediction of user activity during the busy periods. We will investigate those possibilities in our future work.

References

1. Anand, M., Nightingale, E.B., Flinn, J.: Self-tuning wireless network power management. In: Proceedings of the 9th Annual International Conference on Mobile computing and Networking, pp. 176–189. ACM Press, New York (2003),
   http://doi.acm.org/10.1145/938985.939004
2. Anand, M., Nightingale, E.B., Flinn, J.: Ghosts in the machine: Interfaces for better power management. In: Proceedings of the Second International Conference on Mobile Systems, Applications, and Services (MOBISYS 2004) (2004),
   http://www.eecs.umich.edu/~anandm/mobisys.pdf
3. Benini, L., Bogliolo, A., Paleologo, G.A., De Micheli, G.: Policy optimization for dynamic power management. IEEE Transactions on Computer-Aided Design of Integrated Circuits and Systems 18(6), 813–833 (1999)
4. Chung, E.Y., Benini, L., Bogliolo, A., Lu, Y.H., Micheli, G.D.: Dynamic power management for nonstationary service sequests. IEEE Transactions on Computers 51(11), 1345–1361 (2002)
5. Chung, E.Y., Benini, L., Micheli, G.D.: Dynamic power management using adaptive learning tree. In: Proceedings of the International Conference on Computer-Aided Design, pp. 274–279 (1999)
6. Dell Computer Corp.: Dell System 320SLi User's Guide (1992)
7. Ellis, C.S.: The case for higher-level power management. In: Workshop on Hot Topics in Operating Systems, Rio Rico, AZ, USA, pp. 162–167 (1999)
8. Gniady, C., Butt, A.R., Hu, Y.C.: Program counter based pattern classification in buffer caching. In: Proceedings of the 6th Symposium on Operating Systems Design and Implementation (2004)
9. Gniady, C., Hu, Y.C., Lu, Y.H.: Program counter based techniques for dynamic power management. In: Proceedings of the 10th International Symposium on High Performance Computer Architecture (HPCA) (2004)
10. Heath, T., Pinheiro, E., Hom, J., Kremer, U., Bianchini, R.: Application transformations for energy and performance-aware device management. In: Proceedings of the 11th International Conference on Parallel Architectures and Compilation Techniques (2002)
11. Hewlett-Packard: Kittyhawk power management modes. Internal document (1993)
12. Hwang, C.H., Wu, A.C.: A predictive system shutdown method for energy saving of event driven computation. ACM Transactions on Design Automation of Electronic Systems 5(2), 226–241 (2000)
13. Lu, Y.H., Micheli, G.D., Benini, L.: Requester-aware power reduction. In: Proceedings of the International Symposium on System Synthesis, pp. 18–24 (2000)
14. Nightingale, E.B., Flinn, J.: Energy efficiency and storage flexibility in the blue file system. In: Proceedings of the 6th Symposium on Operating Systems Design and Implementation (OSDI) (2004)
15. Qiu, Q., Pedram, M.: Dynamic power management based on continuous-time markov decision processes. In: Proceedings of the Design Automation Conference, New Orleans, LA, USA, pp. 555–561 (1999)
16. Samsung: Samsung Teams with Microsoft to Develop First Hybrid Hard Drive with NAND Flash Memory (2005)

17. Simunic, T., Benini, L., Glynn, P., Micheli, G.D.: Dynamic power management for portable systems. In: Proceedings of the International Conference on Mobile Computing and Networking, pp. 11–19 (2000)
18. Srivastava, M.B., Chandrakasan, A.P., Brodersen, R.W.: Predictive system shutdown and other architecture techniques for energy efficient programmable computation. IEEE Transactions on VLSI Systems 4(1), 42–55 (1996)
19. Weissel, A., Beutel, B., Bellosa, F.: Cooperative I/O—a novel I/O semantics for energy-aware applications. In: Proceedings of the Fifth Symposium on Operating System Design and Implementation (2002)

Effects of Distributed Ordering Mechanism in Task Allocation

Yoshihito Saito and Tokuro Matsuo

Abstract. This paper describes a task allocation mechanism and its performance concerned with software developing. In recent years, multiple types of trading are available since the Internet-based commerce has been developing. Electronic commerce handles items trading including goods, information, times, right, tasks. When tasks are traded, players (software developers) bid each task with the valuation estimated by its cost. Auctioneer (outsourcer) allocates tasks to players based on the order of their valuations and he/she pays money to winners in the auction. However, when a software developer fails to develop the project and becomes bankruptcy, the outsourcer has much amount of damages and the risk of chain bankruptcy. To prevent him/her from such situation, we propose a new contract method based on distributed ordering in which the outsourcer allocates distributed tasks to multiple developers. Winners are determined based on valuations and ability where each developer declares. We also show the result of simulations with the rates of bankruptcy and discuss the effective strategy for outsourcer to allocate the tasks.

1 Introduction

Computer-based commerce is one of effective form of economic activities [1] [2] [3] [4]. Instead of items trading, software can be ordered as tasks from outsourcers to developers electronically. Each trader is implemented as a software/autonomous agent. An outsourcer agent negotiates with developers on costs and quantities of tasks and decides price and allocations based on results of negotiations. He/she or ders tasks to a developer agent. In this paper, we consider the situation that outsourcer orders tasks to many companies in software manufacture. In general,

Yoshihito Saito
Department of Informatics, Graduate School of Engineering, Yamagata University, 4-3-16, Jonan, Yonezawa, Yamagata, 992-8510, Japan
e-mail: saito2007@e-activity.org

Tokuro Matsuo
Department of Informatics, Graduate School of Engineering, Yamagata University, 4-3-16, Jonan, Yonezawa, Yamagata, 992-8510, Japan
e-mail: matsuo@tokuro.net

R. Lee, N. Ishii (Eds.): Software Engineering, Artificial Intelligence, SCI 209, pp. 15–29.
springerlink.com   © Springer-Verlag Berlin Heidelberg 2009

most of large-scale software system consists of multiple modules and sets of classes. If such software is ordered to only one software developing company, it takes a lot of time and costs to complete. If the software system can be divided as some middle sizes of modules, the outsourcer can order the system to multiple developers with divided modules. On the other hand, when there are some risks, such as bankruptcy and dishonor, outsourcers should consider how they can order to developers effectively. These are important to make decisions and strategy to win the competition against other software developers.

In this paper, we analyze the effective strategy for the ordering party to save lots of time and costs. When outsourcer determines developers who can perform task successfully and cheaply. The development company proposes concrete cost of his/her work and the outsourcer ask to discount of the price. Outsourcer company sometimes orders the task as a set to one developer. In this case, he/she considers only one developer's ability to perform the task. However, there is a certain risk since the developer goes out his/her business due to bankruptcy. It takes a lot of money and cost in this situation. Outsourcer needs much money to complete his/her project.

To solve the problem, we consider the divided ordering method avoiding the risk such as chain-reaction bankruptcy. We also consider the model where each task distributes to more developers. Time of development is reduced due to distributed task.

To compare and analyze with the above situations, we give some simulation results for some conditions. Our experiment results show the relationship between the risk of bankruptcy and outsourcer's cost. When number of module increases, outsourcer should order as distribution. In less number of modules and developers, good strategy for outsourcer to reduce cost of ordering is that he/she orders to only one developer. There are less modules in software, outsourcer prevents high costs from the risk. On the other hands, when the number of modules and companies increase more and more, good strategy for outsourcer to reduce cost of ordering is that he/she orders to multiple developers distributionally.

The rest of this paper consists of the following six parts. In Section 2, we show preliminaries on several terms and concepts of auctions. In Section 3, we propose some protocols in distributional software manufacture. In Section 4, we conduct some experiment in situations where number of modules and companies increase. In Section 5, describes some discussions concerned with development of our proposed protocol. Finally, we present our concluding remarks and future work.

2 Preliminaries

2.1 Model

Here, give some definitions and assmptions. The participants of trading consist of an ordering party and multiple software developers. The outsourcer prepares the plan of order to outside manufacturers, and developers declare evaluation values for what they can serve the orders. The outsourcer orders the software development companies to do subcontracted implementing modules. We define that the cost is including developer's all sorts of fee and salary.

- At lease, there is one ordered software project. The architected software consists of a set of dividable module $M = \{m_1,...,m_j,...,m_k\}$. m_j is the jth module in the set.
- d_i is the ith contracted software developer with an outsourcer in a set of developers $D = \{d_1,...,d_i,...,d_n\}$.
- Software developers declare a valuation of work when they can contract in implementation of the modules. $v_{ij}(v_{ij} = 0)$ is the valuation when the developer d_i can contract for implementation of the module m_j.
- P_{ij}^{pre} is an initial payment for developer d_i paid by the outsourcer.
- P_{ij}^{post} is an incentive fee paid after the delivery of the completed module.
- v_{ij} is $P_{ij}^{pre} + P_{ij}^{post}$
- Condition of software development company consists of his/her financial standing, management attitude, firm performance, and several other factors. The condition is shown as A_i integrated by them.
- The set of allocation is $G = \{(G_1,...,G_n) : G_i \cap G_j = \phi, G_i \subseteq G\}$.
- G_i is an allocation of ordering to developer d_i.

Assumption 1 (Number of developers). *Simply, we assume $n > k$. There are lots of software developers.*

Assumption 2 (Payment). *There are two payment such as advanced-initial payment and contingent fee. Realistically, the former increases the incentives of making allocated modules. When the module is delivered successfully, the contingency fee is paid to the developer.*

Assumption 3 (Risks). *In the period of developing the modules, there is a certain risk r_i for developer d_i, such as bankruptcy and dishonor. We assume that r_i is calculate as $1 - A_i$.*

Assumption 4 (Dividable modules). *We assume that the large-scale software can be divided as some middle size modules.*

Assumption 5 (Module's quality). *We assume that the quality of each developed module is equal to the quality of other modules.*

Assumption 6 (Integration of Modules). *We assume that some modules can be integrated without the cost.*

2.2 Initial Payment

When developers serve tasks from ordering company, a partial payment is paid before they start developing modules. In actual unit contract of software implementation, the payment sometimes divided as an initial payment and incentive fee. In this paper, we assume that v_{ij} is $P_{ij}^{pre} + P_{ij}^{post}$. In general, P_{ij}^{pre} is sometimes increased based on the quality of finished work. Simply, we do not consider it.

For example, the value A_i of condition of software development company is calculated based on his/her financial standing, management attitude, firm performance, and several other factors. If A_i is higher value, the developer has the credit. On the other hand, if A_i is near zero, the company does not enough credit. Some companies may have been just now established. If a company has enough credit, they need not do fraud since they have a steady flow of business coming in due to his/her credit.

In this paper, to make simple discussion, we assume that the developers must complete the performance on contract when they are winner of the auction. Concretely, we give the following assumption.

Assumption 7 (Performance on contract). *There are no developers cancel and refusal allocated tasks without performance on contract. Namely, the condition of participation to bidding is performance of business without performance of business.*

2.3 Contract

When an outsourcer orders an architecture of software to development vender company, there are mainly two types of trading. One is the trading by contract at discretion. Ordering company determines the developers ordering making software. They decide the price of the work. For example, the development company proposes concrete cost of his/her work and the outsourcer ask to discount of the price. Another type of trading is a policy of open recruitment. In this trading, the companies who can accept the order from outsourcer compete on price. First, the outsourcer shows the highest value in which they can pay for the scale of software. When the cost for developers is less than the value, they declare to join in the bidding. In this paper, we consider the latter case of contracts.

When developers who participate in the competition, they give bid values for the task. For example, there are three developers d_1, d_2, and d_3. If the developer d_2 bids the lowest value in three developers, he/she contracts for the implementation of software ordered by the outsourcer.

3 Protocol

In this section, we propose concrete protocol to determine the winner of contract. Here, we consider the risk about the developers. There are some risks for developers as a company, such as, bankruptcy and dishonor. To reduce the rate of risks, we propose a new diversification of risk based on divided tasks in large-scale software system manufacture. Further, we employ the advanced-initial payment and contingent fee as payment from an outsourcer. The former increases the incentives of making allocated modules. When the module is delivered successfully, the contingency fee is paid to the developer.

3.1 Protocol 1

We show simple protcl of contract. Figure 1 shows the fllowing contract model. In this figure, developer 2 serves all tasks as a whole. Namely, developer 2 bids the lowest valuation comparing with other all developers. Developer 2 needs to complete all modules.

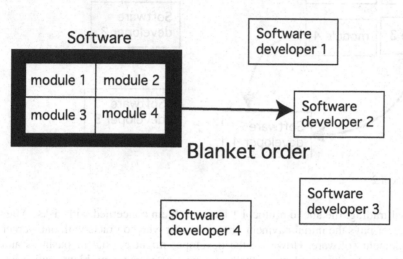

Fig. 1 Protocol 1

1. For the large-scale software, an outsourcer offers for public subscription.
2. Software developers who can contract with the outsourcer come forward as contractor.
3. Developers submit a cost value of the task to the outsourcer by sealed bid auction. Namely, they bid $\sum_{j=1}^{k} m_j$.
4. A developer who bids the $\min \sum_{j=1}^{k} m_j$ contracts with outsourcer for the declared cost.

For example, there are 3 developers. d_1 bids for 100, d_2 bids for 80, and d_3 bids for 50. In this case, developer d_3 serves the work for 50. Here, we consider that the developer 3 go out of business due to a certain factor. We assume that initial fees P_{ij}^{pre} of the work are paid as fifty percent of the contract prices. Namely, incentive fee P_{ij}^{post} is determined another fifty percent of cost. The outsourcer lost initial fee of developer d_3 for 25 dollars. The outsourcer orders and re-allocates the task to the developer d_2 since he/she bids the second lowest value. Totally, the ordering party takes 105 since it needs initial fee for developer d_3 and contract fee of developer d_2. In this contract, it is very high risk for the outsourcer if developer becomes bankruptcy during serving the tasks.

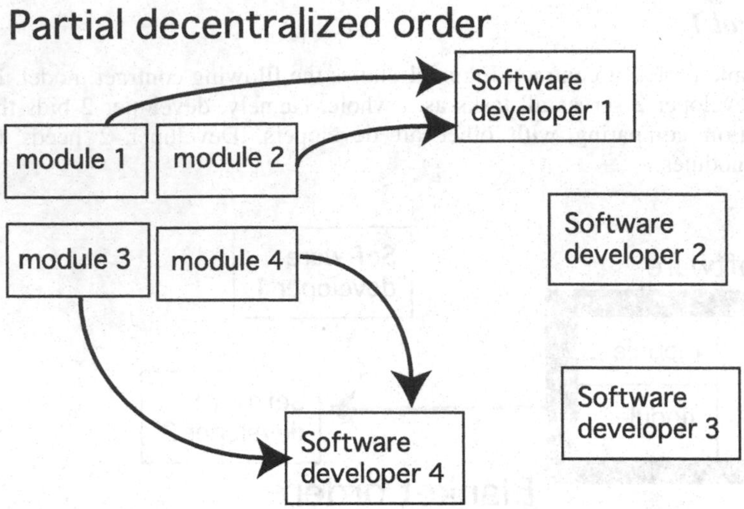

Partial decentralized order

Fig. 2 Protocol 2

In actual trading, the above protocol 1 has a problem concerned with risks. After the outsourcer pays the initial payment, the developer who contracts with outsourcer starts implement software. However, the developer might get out of business and bankruptcy due to the problem of their company's financial problem, and other undesirable factors. When the developer declares his/her cost for five million dollars, the outsourcer pays the initial fee for two million dollars. If the developers become bankruptcy, the outsourcer lost much amount of money. For example, in the auction, a developer who bid for 6 million dollars as the second highest valuation, the outsourcer incrementally takes at least 4 million dollars to complete the software.

To solve the problem, we consider the divided ordering method avoiding the risk such as chain-reaction bankruptcy. We assume the dividable module such as assumption 4. Developers bid their valuations with each module like a combinatorial auction [5] [6]. Figure 2 shows the example of this situation. In this example, developers 1 serves developing module 1 and 2. Developer 4 has the task of development of module 3 and 4. Here, we give a concrete protocol by using the assumption 4.

3.2 Protocol 2

We show a partial task division allocation protocol. Figure 2 shows the following contract model. In this figure, developer 1 serves module 1 and 2, developer 4 serves module 3 and 4. In module 1 and 2, developer 1 bids the lowest valuation comparing with other all developers. In module 3 and 4, developer 4 bids the lowest valuation comparing with other all developers. Each developer needs to complete each modules.

1. For the large-scale software, an outsourcer offers for public subscription. Tasks are divided as multiple modules.
2. Software developers who can contract with the outsourcer come forward as contractor.
3. Developing companies evaluate a value for each module considering the scale of task.
4. Then, they bid their valuations by sealed bid auction. Namely, they bid the set of $\{v_{i1},...,v_{ij}...,v_{ik}\}$.
5. The outsourcer calculates a minimized set of all development parties' valuations. Namely, the outsourcer computes $G = \mathrm{argmin}_i \sum_{j=i}^{k} v_{ij}$.

In this protocol, outsourcer can outsource tasks at the lowest price. The followings are examples of protocol 2.

Example. There are 5 developers. The software consists of 4 modules.

d_1's valuation: $\{v_{11}, v_{12}, v_{13}, v_{14}\}$ is (<u>20</u>, 60, 40, <u>30</u>).
d_2's valuation: $\{v_{21}, v_{22}, v_{23}, v_{24}\}$ is (30, <u>30</u>, 50, 40).
d_3's valuation: $\{v_{31}, v_{32}, v_{33}, v_{34}\}$ is (40, 40, <u>20</u>, 50).
d_4's valuation: $\{v_{41}, v_{42}, v_{43}, v_{44}\}$ is (25, 50, 50, 70).
d_5's valuation: $\{v_{51}, v_{52}, v_{53}, v_{54}\}$ is (50, 40, 60, 60).

Thus, developer d_1 has implementation of module 1 and 4. Developer d_2 serves the work of module 2. Developer d_3 serves the work of module 3. Total costs of ordering company are calculated as $\sum_{i=1}^{4} v_{ij}$ is 110. We assume that initial fees of the work are paid as fifty percent of the contract prices.

Here, we consider that the developer 3 go out of business due to a certain factor. The outsourcer lost initial fee of developer d_3 for ten dollars. The outsourcer orders and re-allocates the task to the developer d_1 since he/she bids the second lowest value. Totally, the ordering party takes 130 dollars since it needs initial fee of module 3 for developer d_3 and contract fee of module 3 with developer d_1.

Realistically, there are some risks in the protocol 2 since the developer 1 might become bankruptcy. Further, it takes much time to complete all tasks since most of tasks sometimes concentrates to one developer. To solve the problem, we consider the model where each task distributes to more developers.

3.3 Protocol 3

We show a task division allocation protocol. Figure 3 shows the example of this situation. In this example, each developer serves one task. Thus, time of development is reduced due to distributed task.

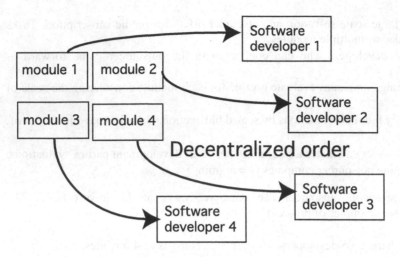

Fig. 3 Protocol 3

1. For the large-scale software, an outsourcer offers for public subscription. Tasks are divided as multiple modules.
2. Software developers who can contract with the outsourcer come forward as contractor.
3. Developing companies evaluate a value for each module considering the scale of task.
4. Then, they bid their valuations by sealed bid auction. Namely, they bid the set of $\{v_{i1}, ..., v_{ij}, ..., v_{ik}\}$.
5. The outsourcer calculates a minimized set of all development parties' valuations. Namely, the outsourcer computes $G = \mathrm{argmin}_i \sum_{j=i}^{k} v_{ij}$ such that each agent serves only one task.

Example. There are 5 developers. The software consists of 4 modules.

d_1's valuation: $\{v_{11}, v_{12}, v_{13}, v_{14}\}$ is (20, 60, 40, <u>30</u>).
d_2's valuation: $\{v_{21}, v_{22}, v_{23}, v_{24}\}$ is (30, <u>30,</u> 50, 40).
d_3's valuation: $\{v_{31}, v_{32}, v_{33}, v_{34}\}$ is (40, 40, <u>20,</u> 50).
d_4's valuation: $\{v_{41}, v_{42}, v_{43}, v_{44}\}$ is (<u>25</u>, 50, 50, 70).
d_5's valuation: $\{v_{51}, v_{52}, v_{53}, v_{54}\}$ is (50, 40, 60, 60).

In this example, the module m_1 is allocated to the developer d_4. Comparing with the previous example, the allocation of module m_1 changes from developer d_1 to d_3. Thus, the tasks are distributed to avoid non-performance on contract. Here, we give one undesirable example when the all tasks are allocated to one company using protocol 2 and we show the effectiveness of protocol 3.

Example. There are 5 developers. The software consists of 4 modules. The protocol 2 is employed.

d_1's valuation: $\{v_{11}, v_{12}, v_{13}, v_{14}\}$ is (50, <u>40</u>, 50, <u>60</u>).
d_2's valuation: $\{v_{21}, v_{22}, v_{23}, v_{24}\}$ is (90, 60, 70, 70).
d_3's valuation: $\{v_{31}, v_{32}, v_{33}, v_{34}\}$ is (70, 70, 60, 80).
d_4's valuation: $\{v_{41}, v_{42}, v_{43}, v_{44}\}$ is (60, 70, 70, 70).
d_5's valuation: $\{v_{51}, v_{52}, v_{53}, v_{54}\}$ is (80, 50, 60, 70).

In this case using protocol 2, all tasks are allocated to developer d_1 since d_1 bids the lowest valuation to all tasks. However, let us consider the following situation. The conditions of software development company are calculated as $\{A_1, A_2, A_3, A_4, A_5\} = \{0.1, 0.9, 0.9, 0.9, 0.9\}$. Namely, the potential rates of bankruptcy of developers are calculated as $\{r_1, r_2, r_3, r_4, r_5\} = \{0.9, 0.1, 0.1, 0.1, 0.001\}$. If the developer d_1 closes his/her business in period of contract, the outsourcer lost $0.1 \cdot (50 + 40 + 50 + 60) = 20$. Additionally, the tasks are re-allocated to remained developer as follows.

d_2's valuation: $\{v_{21}, v_{22}, v_{23}, v_{24}\}$ is (90, 60, 70, <u>70</u>).
d_3's valuation: $\{v_{31}, v_{32}, v_{33}, v_{34}\}$ is (70, 70, <u>60</u>, 80).
d_4's valuation: $\{v_{41}, v_{42}, v_{43}, v_{44}\}$ is (<u>60</u>, 70, 70, 70).
d_5's valuation: $\{v_{51}, v_{52}, v_{53}, v_{54}\}$ is (80, <u>50</u>, 60, 70).

Totally, the outsourcer takes 260 (20 + 240). If protocol 3 is employed, the outsourcer takes 245 (5 + 240) even though the developer d_1 becomes bankruptcy.

4 Periminary Experiments

To compare and analyze the effectiveness of our proposed issues, we conduct simulations concerned with relationships between rate of risks and outsourcer's cost.

Figures 4 and 5 show experimental results where the number of developers and tasks change. We created 100,000 different problems and show the averages of the cost. The vertical axis shows the average cost for outsourcer. The horizontal axis shows the rate of developer's bankruptcy.

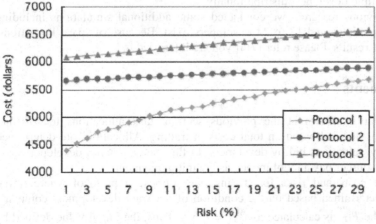

Fig. 4 3 modules and 5 developers

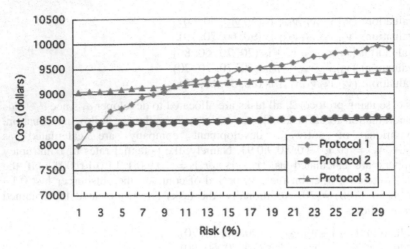

Fig. 5 5 modules and 8 developers

We set the conditions of simulations as follows. Outsourcer orders tasks to developer in prices where developers declare. Namely, the allocations and prices are decided based on sealed first price auction. The cost computation of each task for developers is decided from 1,000 to 4,000 based on uniform distribution. We change the rate of bankruptcy for developers from 1 percent to 30 percent.

Figure 4 shows the result of experiment where the software is divided as 3 modules and 5 developing companies participate in the competition. In this situation, good strategy for outsourcer to reduce cost of ordering is that he/she orders to only one developer. Even though the risks of bankruptcy for developers increase, total cost is less than the protocols 2 and 3.

Figure 5 shows the result of experiment where the software is divided as 5 modules and 8 developing companies participate in the competition. In this situation, good strategy for outsourcer to reduce cost of ordering is that he/she orders to multiple developers distributionally.

In our previous research, we conducted some additional simulations including cases that 8 modules and 12 to 24 developers exist. Because of space limitations, we omit these results. Please refer [7] if you are interested in.

5 Experiments

In the simulation shown at the previous section, task allocations to software developers are decided based on total costs in trading. Allocations are determined based on only valuations bid by developers. In this cases, when a developers gose bankrupt, outsourcer pay a lot of money as the initial payment.

To reduce costs and risks for trading, we propose a protocol where initial payment is determined based on the condition of software development company. We assume the p_{ij}^{pre} is calculated as $p_{ij}^{pre} = r_i \cdot v_{ij}$. Thus, the initial value defined by degree of A_i prevents an outsourcer from intentional bankruptcy by sinister developers. In our protocol, outsourcer order the module to software developer when the initial payment paid to the developer is low. Even thoug, initial payment is low for the developer, he/she cannot cancel serving tasks on our protocol. We conducted some experiments in this situation.

Example. There are 5 developers. The software consists of 4 modules.

d_1's valuation: $\{v_{11}, v_{12}, v_{13}, v_{14}\}$ is (20, 60, 40, 30).
d_2's valuation: $\{v_{21}, v_{22}, v_{23}, v_{24}\}$ is (40, 30, 50, 40).
d_3's valuation: $\{v_{31}, v_{32}, v_{33}, v_{34}\}$ is (30, 40, 20, 50).
d_4's valuation: $\{v_{41}, v_{42}, v_{43}, v_{44}\}$ is (30, 50, 50, 70).
d_5's valuation: $\{v_{51}, v_{52}, v_{53}, v_{54}\}$ is (50, 40, 60, 60).

d_1's condition: $\{A_1\}$ is (0.7).
d_2's condition: $\{A_2\}$ is (0.6).
d_3's condition: $\{A_3\}$ is (0.5).
d_4's condition: $\{A_4\}$ is (0.5).
d_5's condition: $\{A_5\}$ is (0.4).

d_1's initial payment: $\{p_{11}^{pre}, p_{12}^{pre}, p_{13}^{pre}, p_{14}^{pre}\}$ is (14, 42, 28, 21).
d_2's initial payment: $\{p_{21}^{pre}, p_{22}^{pre}, p_{23}^{pre}, p_{24}^{pre}\}$ is (24, 18, 30, 24).
d_3's initial payment: $\{p_{31}^{pre}, p_{32}^{pre}, p_{33}^{pre}, p_{34}^{pre}\}$ is (15, 20, 10, 25).
d_4's initial payment: $\{p_{41}^{pre}, p_{42}^{pre}, p_{43}^{pre}, p_{44}^{pre}\}$ is (15, 25, 25, 35).
d_5's initial payment: $\{p_{51}^{pre}, p_{52}^{pre}, p_{53}^{pre}, p_{54}^{pre}\}$ is (20, 16, 24, 24).

When outsourcer uses protocol 1 by this example, he/she order all to developer d_3. In this case, initial payment is 70, total costs is 140. If d3 goes bankrupt, outsourcer needs to order all module to another, and needs to pay (70 + new order costs). When outsourcer uses protocol 2 by this example, he/she orders module m_1 and m_4 to developer d_1, he/she orders module m_3 to developer d_3, and he/she orders module m_2 to developer d_5. In this case, initial payment is $\{p_{11}^{pre}, p_{52}^{pre}, p_{33}^{pre}, p_{14}^{pre}\}$ =14, 16, 10, 21, total costs is 110. If any developer goes bankrupt, the initial payment that he needs to pay is 35 or less. When outsourcer uses protocol 3 by this example, he/she orders module m_1 to developer d_1, he/she orders module m_4 to developer d_2, he/she orders module m_3 to developer d_3, and he/she orders module m_2 to developer d_5. In this case, initial payment is $\{p_{11}^{pre}, p_{52}^{pre}, p_{33}^{pre}, p_{24}^{pre}\}$ = 14, 16, 10, 24, total costs is 120. Total costs is higher than protocol 2. However, initial payment that outsourcer needs to pay is lower than other protocols.

To compare and analyze the effectiveness of our proposed issues, we conduct simulations on the same condition as Section 4.

Figure 6 shows the result of experiment where the software is divided as 3 modules and 5 developing companies participate in the competition. In this situation, good strategy for outsourcer to reduce cost of ordering is that he/she orders to only one developer. Figure 7 shows the result of experiment where the software is divided as 3 modules and 10 developing companies participate in the competition. In this

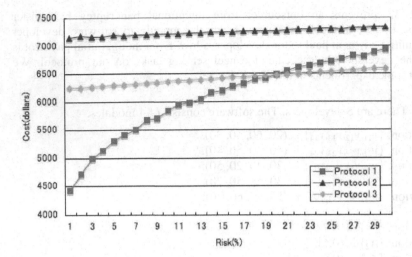

Fig. 6 3 modules and 5 developers

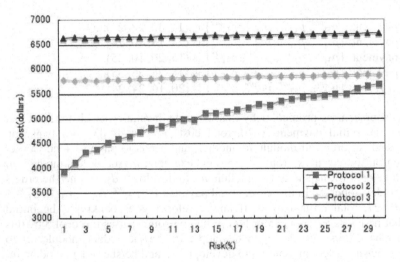

Fig. 7 3 modules and 10 developers

situation, good strategy for outsourcer to reduce cost of ordering is same as the first simulation. In both situations, even though the risks of bankruptcy for developers increase, total costs less than the protocols 2 and 3.

Figure 8 shows the result of experiment where the software is divided as 5 modules and 8 developing companies participate in the competition. This situation only increases two modules and three companies. However, good strategy changes. Good strategy for outsourcer to reduce cost of ordering is that he/she orders each module to each developer. Figure 9 shows the result of experiment where the

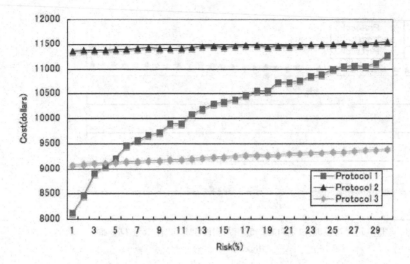

Fig. 8 5 modules and 8 developers

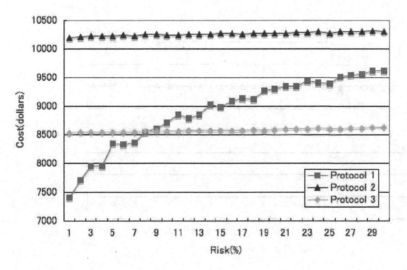

Fig. 9 5 modules and 16 developers

software is divided as 5 modules and 16 developing companies participate in the competition. In this situation, good strategy for outsourcer to reduce cost of ordering is same as the third simulation.

Figure 10 shows the result of experiment where the software is divided as 8 modules and 12 developing companies participate in the competition. Figure 11 shows the result of experiment where the software is divided as 8 modules and 24

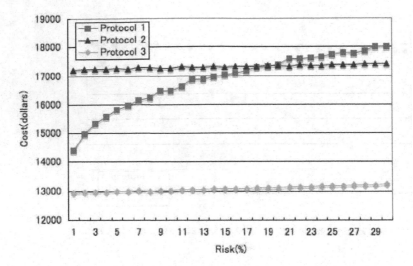

Fig. 10 8 modules and 12 developers

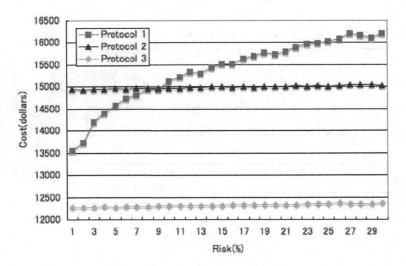

Fig. 11 8 modules and 24 developers

developing companies participate in the competition. In both situation, good strategy for outsourcer to reduce cost of ordering is that he/she orders each module to each developer. In the situation until now, protocol 1 was cheaper than protocol 2. However, in Figure 10 and Figure 11, protocol 1 is higher than protocol 2. If the outsourcer doesn't know in efficient condition in trading. In such situation, the outsourcer pays a lt of costs when he/she may order all modules to one company.

If all modules are ordered to one company when the number of modules is little, the software is developed at a low cost by the developer. When the number of modules is increased, the costs increase if all modules are ordered to oe software developer. When the large-scale software is divided as many modules, the software manufacture is compleate in shout period and at low price.

6 Conclusions

In this paper, we proposed effective strategies for outsourcer to reduce time and cost with number of modules and developers. This means that the outsourcer should not contract with developers at discretion. Further, in auction to determine subcontractors, outsourcer should gather many bidders. Further, ordering party should divide many modules. However, when the number of modules is less, outsourcer should contract only one developer in figure 4.

Our future work includes analysis of situation where each scale of modules is different and analysis of situation where integration of modules takes some costs.

References

1. Matsuo, T., Saito, Y.: Diversification of risk based on divided tasks in large-scale software system manufacture. In: Proc. of the 3rd International Workshop on Data Engineering Issues in E-Commerce and Services (DEECS 2007), pp. 14–25 (2007)
2. Sandholm, T.: Limitations of the vickrey auction in computational multiagent systems. In: Proceedings of the 2nd International Conference on Multiagent Systems (ICMAS 1996), pp. 299–306 (1996)
3. Sandholm, T.: An algorithm for optimal winnr determination in combinatorial auctions. In: Proc. of the 16th International Joint Conference on Artificial Intelligence (IJCAI 1999), pp. 542–547 (1999)
4. Wurman, P.R., Wellman, M.P.: The Michigan internet auctionbot: A configurable auction server for human and software agents. In: Proc. of the 2nd International Conference on Autonomous Agents (AGENT 1998) (1998)
5. Hundson, B., Sandholm, T.: Effectiveness of preference elicitation in combinatorial auctions. In: Proc. of AAMAS 2002 Workshop on Agent Mediated Electronic Commerce IV, AMEC IV (2002)
6. Parkes, D.C., Ungar, L.H.: Iterative combinatorial auctions: Theory and practice. In: Proc. of 17th National Conference on Artificial Intelligence (AAAI 2000), pp. 74–81 (2000)
7. Saito, Y., Matsuo, T.: Analyses of task allocation based on credit constraints. In: Proc. of the 1st International Workshop on Agent-based Complex Automated Negotiation (ACAN 2008) (2008)

Processing of Continuous k Nearest Neighbor Queries in Road Networks

Wei Liao, Xiaoping Wu, Chenghua Yan, and Zhinong Zhong

Abstract. Continuous Nearest Neighbor (NN) monitoring in road networks has recently received many attentions. In many scenarios, there are two kinds of continuous k-NN queries with different semantics. For instance, query "finding the nearest neighbor from me along my moving direction" may return results different from query "finding the nearest neighbor from my current location". However, most existing continuous k-NN monitoring algorithms only support one kind of the above semantic queries. In this paper, we present a novel directional graph model for road networks to simultaneously support these two kinds of continuous k-NN queries by introducing unidirectional network distance and bidirectional network distance metrics. Considering the computational capability of mobile client to locate the edge containing it, we use memory-resident hash table and linear list structures to describe the moving objects and store the directional model. We propose the unidirectional network expansion algorithm and bidirectional network expansion algorithm to reduce the CPU cost of continuous k-NN queries processing. Experimental results show that the above two algorithms outperform existing algorithms including IMA and MKNN algorithms.

1 Introduction

With the advent of mobile computing and massive spread of positioning techniques such as GPS system, location based services have become a promising and challenging area and have gained many attentions in the past few years [1]. Various applications, such as road-side assistance, highway patrol, and location-aware advertisement, are popular in many- especially urban-areas. This has shifted research interests to handle the inherent challenges in designing scalable and efficient architectures to support frequent location updates concurrently. In most urban scenarios, the queries have two distinguished features: (i) using network distance versus traditional Euclidean distance; and (ii) lacking of exact predicted model for

Wei Liao
Dept. of Electronic Engineering, Naval University of Engineering, Wuhan, China
e-mail: liaoweinudt@yahoo.com.cn

Xiaoping Wu, Chenghua Yan, and Zhinong Zhong
School of Electronic Science and Engineering, National University of Defense Technology
Changsha, China

R. Lee, N. Ishii (Eds.): Software Engineering, Artificial Intelligence, SCI 209, pp. 31–42.
springerlink.com   © Springer-Verlag Berlin Heidelberg 2009

moving objects in road network. Therefore, the traditional indexing methods for moving objects and queries processing technologies are not applicable to queries processing in road networks.

Continuous nearest neighbor queries is one of the fundamental problems in the fields of moving objects management [2]. A continuous k-NN query computes the k objects that lie closest to a given query point. Most recent researches have focused on queries processing of moving objects based on an underlying network. Specifically, these techniques aim to continuously monitor a set of moving nearest neighbors. However, one of the limitations of these methods is their reliance on Euclidean distance metrics, which can be imprecise, especially in dense road networks. In general, the main challenges for continuous k-NN processing are to (i) efficiently manage objects location updates and (ii) provide fast network-distance computing. IMA/GMA algorithm [3] is among the few existing technologies applicable for continuous k-NN monitoring, which uses memory-resident architecture to improve the processing efficiency. While MOVNet [4] combines an on-disk R-tree structure to store the connectivity information of the road network with an in-memory grid index to efficiently process moving object position updates.

This paper presents a solution for continuous k-NN queries in road networks, similar to most on-line monitoring systems, we assume main-memory evaluation. Our first contribution is a novel system architecture for continuous k-NN monitoring. We present an enhanced directional graph model for road networks and define two kinds of network metrics: (a) unidirectional network distance and (b) bidirectional network distance to support continuous k-NN queries with different semantics. To facilitate processing of subsequent updates, in-memory hash index and linear list structures are used to manage the current positions of moving objects and store the connectivity information of the road networks. Our second contribution is the *unidirectional network expansion* (UNE) algorithm and *bidirectional network expansion* (BNE) algorithm, using different network metrics to support the above two kinds of continuous k-NN queries and exploiting expansion strategy based on influence tree to reduce the network searching cost of continuous k-NN queries updating.

The rest of the paper is organized as follows. Section 2 reviews related work. Section 3 describes the directional graph model and states the system architecture. In section 4 we propose the UNE and BNE algorithm, and also updating algorithm is discussed. Section 5 experimentally compares the above algorithms through simulations in real road networks. Finally, Section 6 concludes the paper with a summary and directions for future work.

2 Related Work

The existing researches on spatial processing in road networks have been intensively studied recently. Jensen et al. [3] formalized the problem of k-NN search in road networks and presented a system prototype for such queries. Papadias et al. [5] presented an architecture that integrates network and Euclidean information to process network-based queries. This method utilizes the principle that, for any two objects in the network, the network distance is at least the same as the Euclidean distance between them. In contrast, the network expansion strategy performs the query directly from the query point by expanding the nearby vertices in the order of distances from the query point. To avoid on-line distance computation in processing k-NN queries, the VN^3 method [6] was proposed as a Voronoi-based approach to pre-compute the distance. The first NN of any query falling in Voronoi polygon is the corresponding object. If additional NNs are required, the search considers the adjacent Voronoi polygons iteratively. In summary, these above methods hold the assumption that the objects are static.

However, a large number of spatial applications require the capability to process moving objects. SINA [7] executes continuous evaluation of range queries using a three-step spatial join between moving objects and query ranges. Regarding continuous k-NN monitoring on moving objects, there exist three algorithms for exact k-NN queries in the Euclidean space: SEA-CNN [8], YPK-CNN [9], and CPM [10]. All these methods index the data with a regular grid. SEA-CNN performs continuous k-NN queries on moving objects with the idea of sharing execution. Yu et al. proposed YPK-CNN for monitoring continuous k-NN queries by defining a search region based on the maximum distance between the query point and the current locations of previous k nearest neighbors. As an enhancement, Mouratidis et al. presented a solution (CPM) that defines a conceptual partitioning of the space by organizing grid cells in rectangles. However, all these techniques don't consider network distance computation, which makes them unsuitable for network-based applications.

Several works were conducted on continuous k-NN queries in road networks. Mouratidis et al. [3] addressed the issue of processing continuous k-NN queries in road networks by proposing two algorithms (namely, IMA/GMA) that handle arbitrary object and query movement patterns in the road network. This work utilizes an in-memory data structure to store the network connectivity and uses expansion strategy to process concurrently continuous k-NN queries. Wang et al. [4] presented MOVNet architecture and MKNN algorithm for continuous k-NN queries, which adopts a centralized infrastructure with periodical location updates from moving objects. MOVNet uses an on-disk R-tree structure and in-memory grid index to index large graph model of road networks and manage moving objects respectively.

3 System Design

3.1 Assumptions and Network Modeling

In actual applications such as geography information systems and navigation services, the road networks are represented as collection of road nodes and road segments. Specifically road nodes can be the following three cases: (i) the intersections of the road networks, (ii) the dead end of a road segment, and (iii) the points where the curvature of the road segment exceed a certain threshold so that the road segment is split into two to preserve the curvature property. Currently many researches transform the real road networks into in-memory directional graph model to process continuous k-NN queries [3][4]. Specifically, the graph vertexes represent the road nodes, the graph edges represent the road segments, and the weight of each edge represents the road condition or length of road segment. While the moving objects in road networks are denoted as points in graph edges. In many scenarios, moving objects cannot change their directions unless they arrive at the intersection point. For example, in Fig. 1 object o_1 and o_2 moving along the unidirectional road segment n_3n_8, then if o_2 want to catch o_1, it must cross at node n_5 and follow the way n_8n_5, n_5n_2, n_2n_3. while if o_1 want to catch o_2, it only follows along the road segment n_3n_8, obviously the network distance from o_1 to o_2 is not equal to the distance from o_1 to o_2. Specifically, o_2 is the nearest object of o_1, while the nearest object of o_2 is o_4 instead of o_1.

In real applications, users often issue queries with different semantic meanings. As illustrated in Fig. 1, case a) if two moving objects o_1 and o_2 in road segment n_3n_8 issue query "finding the nearest object from me along my moving direction", then with this meaning object o_2 is the nearest object from o_1, while object o_4 is the nearest object from o_2; case b) if o_1 and o_2 issue query "finding the nearest object from my current location", then o_1 and o_2 are their nearest neighbor each other. Recent researches ignored the difference between the two scenarios. To support the above queries, in this paper we propose two kinds of network metrics: unidirectional network distance and bidirectional network distance. Moreover, we transform the real road networks into a unidirectional graph model, converting bidirectional road segment into two unidirectional segments, to support the continuous k-NN monitoring. Fig. 2 demonstrates the corresponding modeling graph of road networks in Fig. 1. The road segment n_2n_5 is represented as unidirectional edge n_2n_5 and n_5n_2. And road segment n_2n_4 is mapped to multiple unidirectional edges. For clarity, we define the graph modeling of road networks as follows.

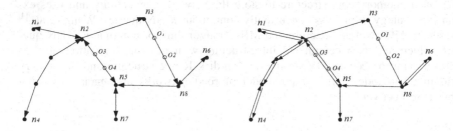

Fig. 1 Real road networks   **Fig. 2** Directional graph model

Definition 1. *A road networks is a directional weighted graph G consisting of a set of edges (road segments) E and a set of vertexes (intersections, deadlines, etc) V, where $E \subseteq V \times V$.*

Definition 2. *For network G, each edge e is represented as $e(v_1, v_2) \in E$, which means it is connected to two vertices v_1 and v_2, where v_1 and v_2 are the starting and ending vertex, respectively. Each edge is associated with a weight, given by a function weight(e): $W : E \to R^+$, where R^+ is the set of positive real numbers.*

The different objects (e.g., cars, taxis, and pedestrians) moving along the road network are identified as the set of moving objects M. the position of a moving object $o \in M$ is represented as $loc (o) = (x, y)$, where x and y are the current location of o, respectively. A query point q is a moving object $\in M$ issuing a location based spatial query at current time. In this paper our design focuses on continuous k-NN queries processing. Note that these queries are processed with network distance.

Definition 3. *The distance of two different objects o_1 and o_2 at time t is $\overrightarrow{dist_t}(o_1,o_2): loc_t(o_1) \times loc_t(o_2) \rightarrow R^+$, $\overrightarrow{dist_t}(o_1,o_2)$ denotes the shortest path from o_1 and o_2 in the metrics of network distance at time t. For simplicity, we denote $\overrightarrow{dist}(o_1,o_2)$ as the distance function of o_1 and o_2 at current time. Usually, $\overrightarrow{dist}(o_1,o_2) \neq \overrightarrow{dist}(o_2,o_1)$.*

Definition 4. *The distance of moving object o and an edge $e(v_1,v_2)$ at time t is defined as $\overrightarrow{dist_t}(o,e): loc(v_1) \times loc_t(o) \rightarrow R^+$, $\overrightarrow{dist_t}(o,e)$ denotes the shortest path from o to starting vertex v_1 of e in the metrics of network distance at time t. for simplicity, we denote $\overrightarrow{dist}(o,e)$ as the distance function of o_1 and o_2 at current time.*

Definition 5. *The $\overrightarrow{unidist_t}(o_1,o_2)$ is the unidirectional distance of object o_1 to o_2 at time t, which is defined as the shortest path $\overrightarrow{dist_t}(o_1,o_2)$ from o_1 to o_2. Usually, $\overrightarrow{unidist_t}(o_1,o_2) \neq \overrightarrow{unidist_t}(o_2,o_1)$. For clarity, we denote $\overrightarrow{unidist}(o_1,o_2)$ as the distance function of o_1 and o_2 at current time.*

Definition 6. *The $\overrightarrow{bidist_t}(o_1,o_2)$ is the bidirectional distance of object o_1 to o_2 at time t, which is defined as the minimum between the shortest path $\overrightarrow{dist_t}(o_1,o_2)$ from o_1 to o_2 and the shortest path $\overrightarrow{dist_t}(o_2,o_1)$ from o_1 to o_2. For short, we denote $\overrightarrow{bidist}(o_1,o_2)$ as the distance function of o_1 and o_2 at current time.*

3.2 Index Structure

Similar with the methods used in [3, 4], our design assumes periodic sampling of the moving objects positions to represent their locations as a function of time. It provides a good approximation on the positions of moving objects. A spatial query submitted by a user at time t_1 is computed based on the locations of moving objects at t_0, that is to say, the system has the last snapshot of moving objects at t_0 with $t_0 \leq t_1$, $t_1 - t_0 < \Delta t$, where Δt is a fixed time interval, and the results are valid until $t_0 + \Delta t$. With the observations that existing mobile devices have sufficient computational capability and memory storage, it can not only locate its current position but also compute the edge it lies in. So at every interval, moving objects submit their current position *loc*, identifier *OID* and within-edge *edgeId* to the server. The identifier *OID* exclusively represents an object, and *edgeId* represent the edge identifier where this object lies in.

Initially, the connectivity and coordinates of nodes in stationary road networks are recorded on disk. For many queries are issued in any possible corner in the road networks. To avoid the frequent retrievals from disk, the following in-memory structures are utilized to represent the modeling graph of road networks. We use a vertex vector named edge table to store all the edges in the modeling graph, a record in edge table is represented as 4-tuple $< edgeId, v_1, v_2, w >$, where *edgeId* denotes the exclusive identifier of an edge in the modeling graph, v_1 and v_2 denote the starting and ending vertexes of this edge, and w denote the weight of this edge. As illustrated in Fig. 2, the record $<n_2n_5, n_2, n_5\ 1>$ represents the edge n_2n_5, where its starting vertex is n_2, ending vertex is n_5, and the length is 1. The second one is a vertex list to store the coordinates of vertices in the graph. For each vertex, a linked list records the edges connecting to it. A record in vertex list is represented as the form $< vId, edgeId_1, edgeId_2, ...label, edgeId_m... >$, where *vId* denotes the identifier of a vertex, and *edgeId* denotes the edges connected to this vertex. The edges before *label* go from this vertex and the edges after *label* end with this vertex. As shown in Fig. 2, $<n_5, n_5n_2, n_5n_7, label, n_2n_5, n_8n_5>$ represents that edge n_2n_5, n_5n_2 go from vertex n_5, and edges n_8n_5, n_5n_7 ends with vertex n_5. The third structure is a hash table used to manage the locations of moving objects, which is represented as the form $< edgeId, o_1, o_2 .... o_n >$, where *edgeId* denotes the identifier of an edge, and $o_i\ (i=1$ to $n)$ denotes the moving objects that lie in the edge. Considering in Fig. 2, $<n_2n_5, o_4>$ represent object o_4 is moving on the edge n_2n_5. With these three structures we can map the moving objects into the modeling graph and process continuous k-NN queries efficiently.

4   Query Design

In this section, we first describe the unidirectional network expansion algorithm and bidirectional network expansion algorithm respectively. And finally we present the updating algorithm based on influence tree.

4.1   Unidirectional Network Expansion (UNE) Algorithm

The main idea of unidirectional network expansion algorithm executes as follows: for each new coming continuous k-NN query, at the periodical time the algorithm first finds the edge that contains the query point. If there exist k moving objects that lie ahead the query point in this edge, then the algorithm return these objects directly; or else the algorithm searches the edges connecting to current edge in the order of unidirectional network distance from the query point to the ending vertex of this edge and prunes edges with current k-NN distance, until finding the k nearest neighbors.

Algorithm 1 details the UNE algorithm. When a continuous k-NN query comes from a query object q, it first gets the edge that contains query object q and pushes this edge into the stack, see Line 2. Then our algorithm fetches the top edge in the stack and search all the objects in this edge, if this edge contains k objects ahead of q, then it returns these k objects (Line 4). Or else the algorithm searches the edge table for adjacent edges to this edge, and pushes them into the stack in order of unidirectional network distance from query q to the ending vertex of this edge (Line 5). Then the algorithm executes the following steps. First, the algorithm fetches the top edge in the stack, gets the k candidate objects in this edge and computes the current k-NN distance $q.kNN.dist$ (Line 6). Then the algorithm pops the top edge and fetches the next edge in the stack, and compares the k-NN distance $q.kNN.dist$ with the distance $\overrightarrow{dist}(q, v_1)$ from query q to starting vertex of current edge. If $q.kNN.dist < \overrightarrow{dist}(q, v_1)$ the algorithm skips this edge, else the algorithm searches current edge for moving objects and update the candidate objects (Line 7 to 12). Note that, the algorithm pushes edges connected to a vertex according to the unidirectional distance from q to the ending vertex of current edge. While the algorithm prunes the edges with the unidirectional distance from q to the starting vertex of current edge.

Algorithm 1. Unidirectional Network Expansion
1. Initialize the empty heap H; set $q.kNN.dist=\infty$
2. Get the edge e that contains q, let e be the root of $q.tree$
3. Insert the m best objects in e into $q.result$; Update $q.kNN.dist$
4. If $m>k$, then return the k NN objects
5. Else search the edges going from $e.end$; en-heap all the edges into H according to its unidirectional distance from q
6. Get k best objects from top edge in H, Update $q.kNN.dist$
7. While $H \neq \phi$, get the next edge e in H
8.   If $\overrightarrow{unidist}(q, e.v_2) < q.kNN.dist$
9.   For each object o in e, Compute the distance between q and o
10.   Update $q.result$, $q.\ kNN.dist$ and $q.tree$
11.   En-heap all the edges going from e into H according its distance from q
12.   Else skip e
13. If the heap $H=\phi$, then return

4.2 Bidirectional Network Expansion (BNE) Algorithm

The steps of BNE algorithm are similar with the UNE algorithm. The difference is that the algorithm must search all the edges connect to this edge when searching the adjacent edges with network expansion strategy.

Algorithm 2 details the BNE algorithm. The algorithm first gets the edge that contains query object q and then searches all the objects in this edge. The algorithm keeps these candidate objects (Line 1 to 4). Note that the algorithm don't return as the final results even there have k best objects, because the BNE algorithm needs expand the network on both directions of current edge, so there maybe objects nearer than these k candidates in adjacent edges. Next, the algorithm searches the edge table for edges adjacent to this edge, and pushes them into the stack in order of bidirectional network distance from query q to the unshared vertex of this edge (Line 5). Then the algorithm executes the following steps (Line 6 to 12): First, the algorithm fetches the top edge in the stack, gets the k candidate objects in this edge and computes the current k-NN bidirectional network distance. Then the algorithm fetches the next edge in the stack, and compares the k-NN bidirectional network distance with the distance from query q to the starting vertex of current edge. If $\overrightarrow{bidist}(q,v_2) < q.kNN.dist$ the algorithm searches current edge for moving objects and updates the candidate objects, else the algorithm skips this edge. Note that, the algorithm pushes edges connected to a vertex according to the bidirectional distance from q to the unshared vertex of current edge. While the algorithm prunes the edges with the bidirectional network distance from q to the shared vertex of current edge.

Algorithm 2. Bidirectional Network Expansion

1. Initialize the empty heap H; set $q.kNN.dist=\infty$
2. Get the edge e that contains q, let e be the root of $q.tree$
3. Insert the m best objects in e into $q.result$;
4. Search edges connecting e, en-heap all the edges into H according to its bidirectional distance from q
5. If $m>k$, Compute $q.kNN.dist$, update $q.result$;
6. Else get k best objects from top edge in H, update $q.kNN.dist$
7. While $H \neq \phi$, get the next edge e in H
8.   If $\overrightarrow{bidist}(q,e.v_2) < q.kNN.dist$
9.   For each object o in e, Compute the bidirectional distance between q and o
10.   Update $q.result$, $q.kNN.dist$ and $q.tree$
11.   En-heap all the edges going from e into H according its distance from q
12.   Else skip e
13. If the heap $H=\phi$, then return

---

5   Experimental Evaluation

5.1   Experimental Settings

To evaluate the performance of BNE and UNE algorithms, we use the Network-based Generator of Moving Objects [11] to generate simulated moving objects. The input is the road map of Oldenburg (a city in Germany) which has 6105 road nodes and 7035 road segments. The generator output 100K objects in each segment of the road networks. Each object randomly chooses a destination. When the object reaches its destination, an update is reported by randomly selecting the next destination. When normalize the data space to 10000×10000, the default velocity of objects is equal to 20 per timestamp. At each timestamp about 0.8% moving objects update their velocities.

We implemented a prototype simulator in C++. The simulation was executed on a workstation with 512MB memory and a 3GHz Pentium4 processor. The simulator first reads node file and edge file of Oldenburg, then constructs in-memory directional graph model. Next, moving objects are read into memory and mapped to corresponding edges. And then the query generator randomly chooses an object and launches a k-NN query from its location. Table 1 summarizes the parameters used. In each experimental setting we vary a single paremeter and keep the others as default values. The experiments evaluate the performance of continuous k-NN queries with mean time of CPU cost (in milliseconds) after 100 iterations.

Table 1 Parameters

Parameters	Default	Value Range
Number of objects (N)	50K	10K-100K
Number of NNs (k)	10	1-25
Objects velocity	Normal	Slow/Normal/High

5.2 Experimental Results

We evaluate the performance of UNE algorithm and BNE algorithm with MKNN algorithm and IMA algorithm in all simulations.

Fig. 3(a) measures the effect of cardinality N (range from 10K to 100K) on the query processing performance when fixed k=10. UNE algorithm outperforms both IMA and MKNN. For MKNN stores the road network on disk and transforms into corresponding modeling graph when needed, thus needs extra disk I/O cost. IMA uses PMR quadtree on the network edges to index moving objects. While our design exploits the hash table to map the moving objects on corresponding edges directly. When an object moves from one edge to another, the mobile device can locate the edge it lies in. So UNE algorithm exhibits a better performance

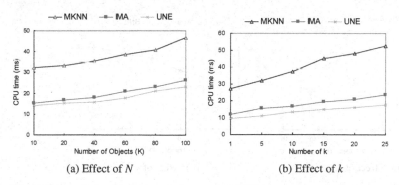

(a) Effect of N (b) Effect of k

Fig. 3 Comparison of querying performance (UNE)

Fig. 3(b) plots the CPU time versus the number of k when fixed the moving objects cardinality N equal to 50K. The performance of UNE gains a slightly improvement than IMA because it can directly access the edge contains query q with an identifier from the mobile device. For UNE and IMA use the in-memory network expansion strategy faster than the on-disk filter-refinement strategy in nature, MKNN shows the worst performance.

Fig. 4(a) illustrates the effect of cardinality N (range from 10K to 100K) on the query processing performance when fixed $k=10$. MKNN has the worst performance for necessary disk I/O when the query q moves outside the influence area. The performance of UNE is better than IMA, for UNE algorithm exploits the influence tree to keep the edges that possible influence the current results and uses network expansion strategy to search the network, thus reducing the CPU time. IMA uses the influence list associated with each edge and expansion tree to support updating of query q and changes of edge weight, thus presents a slightly worse performance than UNE. Fig. 4(b) plots the CPU time versus the number of k when fixed the moving objects cardinality N equal to 50K. It is clear that the CPU time of UNE, IMA and MKNN increase linearly with the number of k, while UNE algorithm shows a best performance.

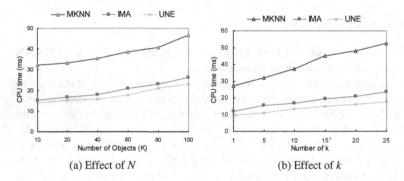

(a) Effect of N (b) Effect of k

Fig. 4 Comparison of updating performance (UNE)

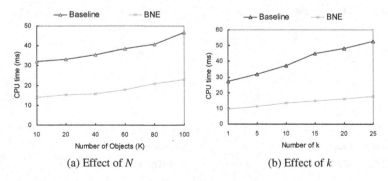

(a) Effect of N (b) Effect of k

Fig. 5 Comparison of querying performance (BNE)

For there is no existing system targeting query processing based on bidirectional distance, we leveraged the concept of network expansion to design *baseline* algorithm without hash table and influence tree structures for performance comparison with BNE algorithm. Fig. 5 (a) shows the effect of cardinality N (range from 10K to 100K) on the query processing performance when fixed $k=10$. BNE outperforms the *baseline* algorithm greatly, while the performance degrade with a similar extent, for our design uses the hash table to identify the edge contain query q, thus reducing greatly searching cost to locate the edge. Fig. 5(b) plots the CPU time versus the number of k when fixed the moving objects cardinality N equal to 50K. The CPU cost of BNE and *baseline* algorithm degrade slowly with the growth of k, for the searching edges ranges from about one edge to a little more than ten edges when k ranging from 1 to 25.

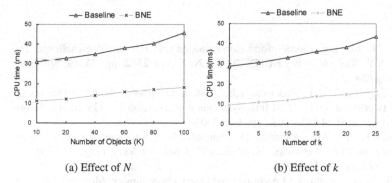

| (a) Effect of N | (b) Effect of k |

Fig. 6 Comparison of updating performance (BNE)

Fig. 6 (a) shows the effect of cardinality N (range from 10K to 100K) on the updating performance when fixed $k=10$. BNE outperforms the *baseline* algorithm greatly, for BNE uses the influence tree to keep the last searching region, thus reducing the edge searching cost. Fig. 6(b) plots the CPU time versus the number of k when fixed the moving objects cardinality N equal to 50K. The CPU cost of BNE and *baseline* algorithm degrade slowly with the growth of k, while BNE shows an excellent performance in each conditions.

6 Conclusions

This paper contributes the first work supporting unidirectional network distance based query processing. We present a novel graph model, supporting two kinds of k-NN query processing with different semantics. We define two kinds of network measurements (a) unidirectional network distance and (b) bidirectional network distance. To facilitate processing of subsequent updates, in-memory hash index and linear list structures are used to manage the current positions of moving objects and store the connectivity information of the road networks. We propose the *unidirectional network expansion* (UNE) algorithm and *bidirectional network expansion* (BNE) algorithm, using network metrics and expansion strategy to reduce the network searching cost of continuous k-NN queries processing.

Our work aims at minimizing the CPU cost at the central processing server. An interesting direction for future work is to extend our design on multi-core processors, and design a scalable and parallel algorithm for many concurrent user scenarios. Another challenging research direction is monitoring of different queries in road network that have counterparts in Euclidean distance such as reverse NN queries and group NN queries.

Acknowledgments. This work is supported by the National High Technology Development Plan under grant No. 2007AA12Z208 and China Postdoctoral Science Foundation under grant No.20080431384.

References

[1] Wolfson, O.: Moving objects information management: The database challenge. In: Halevy, A.Y., Gal, A. (eds.) NGITS 2002. LNCS, vol. 2382, pp. 75–89. Springer, Heidelberg (2002)

[2] Jensen, C.S., Kolárvr, J., Pedersen, T.B., Timko, I.: Nearest Neighbor Queries in Road Networks. In: Proc. ACM Intl. Symposium on Advances in Geographic Information Systems(ACM GIS 2003), pp. 1–8 (2003)

[3] Mouratidis, K., Yiu, M.L., Papadias, D., Mamoulis, N.: Continuous Nearest Neighbor Monitoring in Road Networks. In: Proc. Intl. Conf. on Very Large Data Bases (VLDB 2006), pp. 43–54 (2006)

[4] Wang, H., Zimmermann, R.: Location-based Query Processing on Moving Objects in Road Networks. In: Proc. Intl. Conf. on Very Large Data Bases (VLDB 2007), pp. 321–332 (2007)

[5] Papadias, D., Zhang, J., Mamoulis, N., Tao, Y.: Query Processing in Spatial Network Databases. In: Proc. Intl. Conf. on Very Large Data Bases, pp. 802–813 (2003)

[6] Kolahdouzan, M.R., Shahabi, C.: Voronoi-Based K Nearest Neighbor Search for Spatial Network Databases. In: Proc. Intl. Conf. on Very Large Data Bases (VLDB 2004), pp. 840–851 (2004)

[7] Mokbel, M.F., Xiong, X., Aref, W.G.: SINA: Scalable Incremental Processing of Continuous Queries in Spatiotemporal Databases. In: Proc. Intl. Conf. on Management of Data (SIGMOD 2004), pp. 623–634 (2004)

[8] Xiong, X., Mokbel, M.F., Aref, W.G.: SEA-CNN: Scalable Processing of Continuous K-Nearest Neighbor Queries in Spatio-temporal Databases. In: Proc. Intl. Conf. on Data Engineering (ICDE 2005), pp. 643–654 (2005)

[9] Yu, X., Pu, K.Q., Koudas, N.: Mointoring k-Nearest Neighbour Queries over Moving Objects. In: Proc. Intl. Conf. on Data Engineering (ICDE 2005), pp. 631–642 (2005)

[10] Mouratidis, K., Hadjieleftheriou, M., Papadias, D.: Conceptual Partitioning: An Efficient Method for Continuous Nearest Neighbor Monitoring. In: Proc. Intl. Conf. on Management of Data, Baltimore (SIGMOD 2005), pp. 634–645 (2005)

[11] Brinkhoff, T.: A Framework for Generating Network based Moving Objects. GeoInformatica 6, 153–180 (2002)

A Self-adaptive Greedy Scheduling Scheme for a Multi-Objective Optimization on Identical Parallel Machines

Liya Fan, Fa Zhang, Gongming Wang, Bo Yuan, and Zhiyong Liu

Abstract. A self-adaptive greedy scheduling scheme is presented to solve a Multi-Objective Optimization on Identical Parallel Machines. The primary objective is to minimize the makespan, while the secondary objective makes the schedule more stable. Actual experiments revealed that the scheme obtained the optimal primary and secondary objectives for most cases. Moreover, schedules produced by the scheme were more robust, with smaller makespans. Additionally, it has been applied to parallelize one major component of EMAN, one of the most popular software packages for cryo-electron microscopy single particle reconstruction. Besides, it can also be used in practice to parallelize other similar applications.

Keywords: Multi-objective programming; Load balancing; Scheduling scheme; Parallel computing.

1 Introduction

Recently, cryo-electron microscopy has been used in a great number of applications [1]. For many of them, the key is determining 3-D structures of protein molecules. Traditional techniques, like X-ray crystallography and nuclear magnetic resonance (NMR), have been used to determine structures of protein molecules. However, there are too many limitations for them [2]. As a result, more attention has been given to the cryo-electron microscopy single particle reconstruction technology. Now it has been regarded as a new powerful way to determine 3-D structures of macromolecules.

Liya Fan, Fa Zhang, Gongming Wang, and Zhiyong Liu
Institute of Computing Technology, Chinese Academy of Sciences, Beijing, China
fanliya@ict.ac.cn, zf@ncic.ac.cn, wanggongming@ict.ac.cn,
zyliu@ict.ac.cn

Liya Fan and Gongming Wang
Graduate University of Chinese Academy of Sciences, Beijing, China

Bo Yuan
Department of Computer, Shanghai Jiao Tong University, Shanghai, China
yuanbo@cs.sjtu.edu.cn

R. Lee, N. Ishii (Eds.): Software Engineering, Artificial Intelligence, SCI 209, pp. 43–55.
springerlink.com   © Springer-Verlag Berlin Heidelberg 2009

Several software packages have been widely used for single particle reconstruction [2, 3, 4], and recent advances have made it possible to determine 3-D structures of macromolecules at near-atomic resolutions(~3.8-4.5Å) [5]. However, one of the major problems with these packages is that they are too time-consuming [6]. Unfortunately, only a little work [7] has been done to parallelize these software packages.

Among all these software packages, EMAN is one that achieves the highest resolutions. Its reconstruction process consists of three basic steps: first, a large number of particle images are picked from the raw micrographs; second, an initial 3-D model is generated; third, initial model of the second step is refined iteratively [2]. The third step, which is carried out by the refine program, takes most of the time. It often takes weeks even months to refine a structure.

There are four basic steps of the refine program, projections generation, classification, classes aligning, and 3-D model generation. Classification and classes aligning are the two most time-consuming steps. In our experiments, these two steps took up more than 95% of the total time. Moreover, classes aligning is the step that is most difficult to parallelize, because the problem of load balancing reduces to a multi-objective optimization. Consequently, parallelizing this step is the key to parallelizing EMAN.

In this study, a self-adaptive greedy scheduling scheme is introduced to solve the core problem of parallelizing the classes aligning step, a Multi-Objective Optimization on Identical Parallel Machines. Experiments showed that the scheduling scheme outperformed other existing scheduling algorithms, with a trivial execution time.

2 Problem Description

Four steps of the *refine* program are closely related. During the classification step, N selected particle images are clustered into n groups by the *classesbymra* program, with group i containing k_i images. Then, in the classes aligning step, a class average is generated by the *classalign2* program for each image group.

According to the traditional method of task partitioning, k_i images of group i are subdivided, and assigned to different processors. This method is not efficient here, because the numbers of images in different groups vary greatly. For a group with a small number of images, subdividing the images may result in degraded performance due to the costs incurred by multi-thread and multi-process.

Our solution is to treat each execution of *classalign2* for one group as an independent task. Thus the problem is equivalent to finding a schedule $s: J \rightarrow P$, to distribute tasks to processors. Here $J = \{1, 2... n\}$ contains n independent tasks, and $P = \{1, 2 ... m\}$ contains m identical processors. For convenience, we denote jobs and processors by their indices. The schedule should achieve as balanced loads as possible. Suppose the processing time of task j is p_j, the problem can be formulated as:

(MMIPM) Minimize p

subject to $\sum_{j=1}^{n} p_j x_{ij} \leq p \quad i = 1, 2 \ldots m$

$\qquad \sum_{i=1}^{m} x_{ij} = 1 \quad j = 1, 2 \ldots n$

$\qquad x_{ij} \in \{0,1\} \quad i = 1, 2 \ldots m \quad j = 1, 2 \ldots n$

This is a problem of Makespan Minimization on Identical Parallel Machines (MMIPM), an NP-hard optimization problem [8].

In reality, the problem we face is even more challenging. The processing time is not known before the task is executed, so the schedule has to be constructed on the basis of an estimation of processing times. More often than not, the estimated and actual processing times are not exactly the same, which may affect the balancing of loads produced by the schedule. To account for such differences, the processing time of each task can be considered as a random variable, so is the finishing time of each processor. If we consider the estimated processing time p_j as the mean of the processing time of task j, then the process of solving MMIPM is trying to obtain a schedule that equalizes the means of processors' finishing times. In addition, to make the schedule more stable, it is also desirable that the variances of different processors' finishing times be equal.

Without loss of generality, we suppose that processing times of different tasks have the same variance σ^2, and are independent of each other. Then the variance of each processor's finishing time is $\sigma^2 \sum_{j=1}^{n} x_{ij} \ (i = 1, 2 \ldots m)$, which depends on the number of tasks assigned to the processor. Therefore, in order to equalize variances of different processors' finishing times, each processor should be assigned almost the same number of tasks. Hence, the problem turns out to be a Multi-Objective Optimization on Identical Parallel Machines (MOOIPM). The primary objective minimizes the time of processing all tasks. The secondary objective makes the schedule more stable and robust.

(MOOIPM) Minimize $[p, q]$

subject to $\sum_{j=1}^{n} p_j x_{ij} \leq p \quad i = 1, 2 \ldots m$

$\qquad \sum_{j=1}^{n} x_{ij} \leq q \quad i = 1, 2 \ldots m$

$\qquad \sum_{i=1}^{m} x_{ij} = 1 \quad j = 1, 2 \ldots n$

$\qquad x_{ij} \in \{0,1\} \quad i = 1, 2 \ldots m \quad j = 1, 2 \ldots n$

3 Related Work

The algorithm of solving MOOIPM can be based on the algorithm of solving MMIPM. The most common way to address an NP-hard optimization problem is to construct an approximation algorithm. In 1966, R.L. Graham introduced the first 2-approximation algorithm [9] for MMIPM. Three years later, he presented another 4/3 approximation algorithm [10]. These algorithms are simple and easy to implement. But in our experiments, the loads produced by them were not balanced enough, and even a little load unbalance could result in much extra execution time, because execution of *classalign2* often lasts for a long time.

Some polynomial approximation schemes have also been presented. The first one was presented by Graham in 1969 [10]. Sahni gave another one in 1976 [11]. Unfortunately, time complexities of these schemes are exponential in the number of processors. The scheme introduced by Hochbaum and Shmoys in 1987 [12] overcame such shortcoming. But it is rather difficult to implement. Besides, its huge time complexity $O(\log \frac{1}{\varepsilon}(\frac{n}{\varepsilon})^{\frac{1}{\varepsilon}\log\frac{1}{\varepsilon}})$ makes it unacceptable for our case. The decision problem of MMIPM is strongly NP-hard [13], so there cannot be a fully polynomial approximation scheme unless P = NP **[14]**.

4 The Scheduling Scheme

Our scheduling scheme is organized in three layers. Each layer has its specific responsibilities.

4.1 The Self-adaptive Framework

The first problem is to determine processing times of tasks. The time complexity of *classalign2* is linear to the number of images, so we can model the processing time as a linear function:

$$p_j = ak_j + b, \quad j = 1, 2...n \tag{4-1}$$

where k_j is the number of images in group j, and a and b are coefficients whose values can be obtained by means of the least square method.

The *refine* program is executed in an iterative manner. For n groups of images, *classalign2* is called n times by *refine* in each round. In practice, initial values of (4-1) can be chosen as $a = 1, b = 0$. During each round, first, all tasks are dispatched to processors on the basis of estimated processing times p_j. Second, *classalign2* is invoked for each task and the actual processing time is recorded. Next, values of a and b are updated according to actual processing times. Finally, processing times for the next round is estimated. This process is implemented by the *SADS* (Self-Adaptive Dynamic Scheduling) algorithm:

SADS
1. Set initial values, $a^{(0)} = 1$, $b^{(0)} = 0$, and $p_j^{(0)} = k_j^{(0)}$, $j=1,2...n$
2. For $k=0$ to $ITER - 1$ do
    2.1 Distribute n tasks to m processors by means of estimated values of $p_j^{(k)}$.
    2.2 Run classalign2 for each task and record the actual processing time $t_j^{(k)}$.
    2.3 Update values of a and b by means of the least square method:

$$a^{(k+1)} = \frac{n\sum_{i=1}^{n} k_i^{(k)} t_i^{(k)} - \sum_{i=1}^{n} k_i^{(k)} \sum_{i=1}^{n} t_i^{(k)}}{n\sum_{i=1}^{n} (k_i^{(k)})^2 - (\sum_{i=1}^{n} k_i^{(k)})^2}$$

$$b^{(k+1)} = \frac{\sum_{i=1}^{n} t_i^{(k)} \sum_{i=1}^{n} (k_i^{(k)})^2 - \sum_{i=1}^{n} k_i^{(k)} \sum_{i=1}^{n} k_i^{(k)} t_i^{(k)}}{n\sum_{i=1}^{n} (k_i^{(k)})^2 - (\sum_{i=1}^{n} k_i^{(k)})^2}$$

    2.4 Estimate processing times of the next round:
$$p_j^{(k+1)} = a^{(k+1)} k_j^{(k+1)} + b^{(k+1)}, \quad j=1,2...n$$

End

Obviously, this is a self-adaptive algorithm. In step 2.1, in order to dispatch n tasks to m processors, the *MOS* (Multi-Objective Search) algorithm is called, and the details will be discussed in the next section.

4.2 The Greedy Strategy

This section describes how to dispatch tasks to processors by means of a greedy strategy. Let s_i be a set of tasks that will be dispatched to i processors, and $f_i(s_i)$ be the minimum makespan. Let $g_i(s_i)$ be the largest number of tasks for any of the i processors. The following dynamic programming model can be used to obtain the efficient solution of MOOIPM:

$$f_i(s_i) = \min_{u \subseteq s_i}(\max(f_{i-1}(s_i - u), \sum_{j \in u_i} p_j)) \qquad (4\text{-}2)$$

$$U_i = \{u \mid u \subseteq s_i, \max(f_{i-1}(s_i - u), \sum_{j \in u} p_j) = f_i(s_i)\} \qquad (4\text{-}3)$$

$$g_i(s_i) = \min_{u_i \in U_i}\{\max\{g_{i-1}(s_{i-1}), |u_i|\}\} \qquad (4\text{-}4)$$

$$s_{i-1} = s_i - u_i \qquad (4\text{-}5)$$

$$f_1(s_1) = \sum_{j \in s_1} p_j \qquad (4\text{-}6)$$

$$g_1(s_1) = |s_1| \qquad (4\text{-}7)$$

U_i contains all subsets of s_i that yield the minimum makespan, and u_i represents tasks assigned to processor i. Directly solving this model is impractical. For example, evaluating $f_i(s_i)$ requires trying all 2^{u_i} possible values of u.

According to our greedy strategy, to get the efficient solution, all processors should have almost the same finishing time, and almost the same number of tasks. In other words, the finishing time of each processor should be as close to $\frac{1}{i}\sum_{j\in s_i} p_j$ as possible, and the number of tasks should be as close to $|s_i| / i$ as possible. This can be formulated as a Multi-Objective Integer Program (MOIP).

(MOIP)

Minimize $[\left| \frac{1}{i}\sum_{j\in s_i} p_j - \sum_{j\in s_i} p_j y_j \right|, \left| \frac{|s_i|}{i} - \sum_{j\in s_i} y_j \right|]$

subject to $y_j \in \{0,1\}\quad j\in s_i$

By focusing on the primary objective, MOIP can be reduced to two Multi-Objective Integer Programs:

(MOIP1)

Minimize $[\frac{1}{i}\sum_{j\in s_i} p_j - \sum_{j\in s_i} p_j y_j , \left| \frac{|s_i|}{i} - \sum_{j\in s_i} y_j \right|]$

subject to $\sum_{j\in s_i} p_j y_j \leq \frac{1}{i}\sum_{j\in s_i} p_j$

$y_j \in \{0,1\}\quad j\in s_i$

(MOIP2)

Minimize $[\sum_{j\in s_i} p_j y_j - \frac{1}{i}\sum_{j\in s_i} p_j , \left| \frac{|s_i|}{i} - \sum_{j\in s_i} y_j \right|]$

subject to $\sum_{j\in s_i} p_j y_j \geq \frac{1}{i}\sum_{j\in s_i} p_j$

$y_j \in \{0,1\}\quad j\in s_i$

If we temporarily eliminate the second objective of MOIP1 and MOIP2, it can be noted that MOIP1 becomes a subset-sum problem [15], and MOIP2 can be easily transformed to a subset-sum problem. Therefore, solving MOIP1 and MOIP2 can be based on the algorithm for the subset-sum problem. The details will be described in the next section. Given the algorithm to solve MOIP, the dynamic programming model can be adopted to solve the problem as follows:

MOS($\{p_1, p_2 \ldots p_n\}, m, n$)

1. Set $s \leftarrow \{1, 2 \ldots n\}$, $p \leftarrow \frac{1}{m}\sum_{j=1}^{n} p_j$, $q \leftarrow \frac{n}{m}$

2. for $i = 1$ to m-1 do
    2.1 $P \leftarrow \{p_j \mid j\in s\}$, $u_i = $ DMOSS (P, p, q)

$$2.2 \quad f_i = \max(f_{i-1}, \sum_{j \in u_i} p_j), \quad g_i = \max(g_{i-1}, |u_i|)$$

$$2.3 \quad s \leftarrow s - u_i, \quad p \leftarrow \frac{1}{m-i} \sum_{j \in s} p_j, \quad q \leftarrow \frac{|s|}{m-i}$$

end

$$3. \quad u_m = s, \quad f_m = \max(f_{m-1}, \sum_{j \in s} p_j), \quad g_m = \max(g_{m-1}, |s|)$$

Input arguments m and n are the numbers of processors and tasks, respectively. p_j is the processing time of task j. The algorithm *DMOSS* (Dual Multi-Objective Subset-Sum) of step 2.1 is used to solve MOIP, and will be discussed in the next section.

4.3 The Algorithm to Solve MOIP

Solving MOIP can be based on the algorithm for the subset-sum problem; however, the decision problem of subset-sum is also NP-hard [15]. Taking into account of the specific situation, we believe that some reasonable simplifications can be made to MOIP. First, it can be easily proved that the target value of the primary objective of MOIP ($\frac{1}{i} \sum_{j \in s_i} p_j$) has an upper bound polynomial in the input size:

$$\frac{1}{i} \sum_{j \in s_i} p_j \le \sum_{j \in s_i} p_j \le \sum_{j=1}^{n} p_j = \sum_{j=1}^{n} (ak_j + b) = aN + nb \tag{4-8}$$

where N is the total number of particle images; a and b are constant coefficients. Another simplification is that processing times p_j ($j = 1, 2 \dots n$), should be multiples of some constant c. This can be achieved by rounding each p_j to the nearest multiple of c. The value of c can be adjusted according to specific situations.

Given these simplifications, the primary objective of MOIP can be achieved in polynomial time. Our algorithm is based on the exact algorithm of the subset-sum problem [15]. For other problems, where the first simplification does not apply, the polynomial time approximation scheme for the subset-sum problem [15] can be adopted. We modified the algorithm [15] so that there is no need to run twice for MOIP1 and MOIP2 respectively. The framework of the algorithm is shown below, the input of which are processing times of tasks that are not dispatched yet, as well as target values for the primary and secondary objectives of MOIP.

```
DMOSS ( { p1, p2 ... pr }, p, q)
1. Set S0 = {<0,0>}
2. for i = 1 to r do
    2.1       Si = Merge(Si-1, Si-1 +* <pi, 1>)
    2.2       Truncate(Si, p)
    2.3       UpdateScore(Si, p, q)
end
3. Find the tuple <s, k> from Sr, so that |s - p| is the minimal.
4. Back trace from <s, k> to get the solution.
```

The secondary objective of MOIP is achieved by means of our heuristics. To apply the heuristics, more information is required to be recorded. As a result, elements of S_i $(i = 0, 1 \ldots r)$ are 2-tuples. The first element of each tuple in S_i represents the sum produced by some values from $\{p_1, p_2 \ldots p_i\}$, and the second element represents the number of values that add up to that sum.

The general idea of the heuristics is to give a score to each tuple. The score is based on the difference between the expected and actual numbers of tasks. Consequently, when tuples have identical first element, the one with a smaller score is preferred. For any tuple $<s, k>$ in S_i, the following formulas can be applied to calculate the score:

$$score(s,k) = \left| \frac{i}{r} \cdot q - k \right| \qquad (4\text{-}9)$$

$$score(s,k) = \left| \frac{s}{p} \cdot q - k \right| \qquad (4\text{-}10)$$

The actual number of tasks is k in both formulas. The expected number of tasks is evaluated by the current round number and the current partial sum, respectively. Combining these two formulas leads to hybrid approaches, as illustrated by (4-11) and (4-12), where α and β are constant coefficients.

$$score(s,k) = \left| \alpha(\frac{i}{r} \cdot q - k) + \beta(\frac{s}{t} \cdot q - k) \right| \qquad (4\text{-}11)$$

$$score(s,k) = \alpha \left| \frac{i}{r} \cdot q - k \right| + \beta \left| \frac{s}{t} \cdot q - k \right| \qquad (4\text{-}12)$$

The *Merge* function of step 2.1 is used to merge two sets. If the first elements of two tuples are identical, only the one with the smaller score is retained Operation "+*" of step 2.1 is defined element-wise:

$$S +* < a, b >= \{< x + a, y + b >|< x, y >\in S\} \qquad (4\text{-}13)$$

In step 2.2, the function *Truncate* removes all tuples whose first elements are greater than or equal to p, except the one with the smallest first element greater than or equal to p. *UpdateScore* function of step 2.3 calculates new scores for tuples in the set. Step 4 traces back from S_r to S_1 to get the solution for MOIP.

4.4 Analysis of the Scheme

It can be observed that time complexity of *DMOSS* depends on the loop of step 2, which iterates r times. Step 2.1 can be in $O(|S_{i-1}|)$. Steps 2.2 and 2.3 both require time no more than $O(|S_i|)$. Therefore, each iteration of step 2 takes time $O(|S_i|)$. Since $|S_i| \leq \lceil p/c \rceil + 1$, the total time for step 2 is $O(rp/c)$, so is the time complexity of *DMOSS*

Similarly, time complexity of *MOS* depends on the loop of step 2, which in turn depends on step 2.1. The time of step 2.1 is $O(r^{(i)}p^{(i)}/c)$, where $r^{(i)}$ and $p^{(i)}$ are the number of tasks and target value for the *i*th round of iteration, respectively. By observing $r^{(i)} \leq n$ and formula (4-8), the following formula holds:

$$\frac{r^{(i)}p^{(i)}}{c} \leq n(\frac{a}{c}N + \frac{b}{c}n)$$

Therefore, for the worst case, one iteration of step 2 takes $O(nN + n^2)$ time, and the total time for *MOS* is $O(mnN + mn^2)$. In practice, $N = \alpha n$ often holds, where α is a constant, so the time complexity of *MOS* in the worst case is $O(mn^2)$. For the average case, processors have almost the same finishing time, so the target value of the primary objective is always around

$$\frac{1}{m}\sum_{j=1}^{n} p_j = \frac{1}{m}(aN + bn)$$

Therefore the time for each iteration is $O(nN/m + n^2/m)$, and time complexity for *MOS* is $O(nN + n^2) = O(n^2)$. The average-case time complexity of *MOS* is independent of the number of processors.

Finally, the time complexity of *SADS* is $O(mn^2 \cdot ITER)$ for the worst case, and $O(n^2 \cdot ITER)$ for average case, where *ITER* is the number of iterations.

5 Experimental Results

We implemented the scheme and tested on DAWNING 4000 cluster. The experimental results are shown from two aspects. First, *MOS* is applied to a random dataset and compared with other scheduling algorithms. Second, *SADS* was testified in an actual process of single particle reconstruction.

5.1 Results of a Random Dataset

The random dataset has 120 integers representing processing times of 120 tasks. It was used as the input of MOOIPM. Seven algorithms were tested with different numbers of processors. *Greedy* represents the greedy scheduling algorithm of [9]. *ImpGrd* represents the improved greedy algorithm of [10]. *NoScore* represents *MOS* without scoring heuristics. *Heuristic1*, *Heustic2*, *Hybrid1* and *Hybrid2* are algorithms of *MOS* with scoring formulas (4-9), (4-10), (4-11) and (4-12), respectively. For *Hybrid1* and *Hybrid2*, values of α and β were set to 0.5. Rounding constant was chosen to be 1 for variants of *MOS* (*NoScore, Heuristic1, Heuristic2, Hybrid1 and Hybrid2*).

Table 1 shows the primary and secondary objectives obtained by different algorithms. The bold fold numbers are the optimal values. It can be noticed that variants of *MOS* produced better primary objectives than *Greedy* and *ImpGrd* for all the cases, and for most cases, variants of *MOS* achieved the optimal primary objectives. *Hybrid1* achieved all the optimal primary objectives. In addition, *MOS* with heuristics (*Heuristic1, Heuristic2, Hybrid1 and Hybrid2*) achieved the optimal secondary objectives most of the cases, which meant that our heuristics worked. Finally, the advantages of our scheduling scheme become more notable as the number of processors increases.

Table 1 Values of primary and secondary objectives

#processor		greedy	ImpGrd	NoScore	Heuristic1	Heuristic2	Hybrid1	Hybrid2
4	Primary	75323	74861	**74858**	**74858**	**74858**	**74858**	**74858**
	Secondary	31	**30**	36	**30**	**30**	**30**	**30**
6	Primary	51089	49939	**49905**	**49905**	**49905**	**49905**	**49905**
	Secondary	23	**20**	25	**20**	**20**	**20**	**20**
8	Primary	39093	37473	**37429**	**37429**	**37429**	**37429**	**37429**
	Secondary	18	16	21	16	**15**	**15**	**15**
10	Primary	31443	29969	**29943**	**29943**	**29943**	**29943**	**29943**
	Secondary	15	13	15	**12**	**12**	**12**	**12**
12	Primary	26966	25028	**24953**	**24953**	**24953**	**24953**	**24953**
	Secondary	15	11	12	**10**	**10**	**10**	**10**
16	Primary	20013	18809	**18715**	**18715**	**18715**	**18715**	**18715**
	Secondary	13	9	10	**8**	**8**	9	**8**
20	Primary	16734	15093	14973	14975	14973	**14972**	14976
	Secondary	10	7	7	7	7	7	**6**

Table 2 Variances of finishing times and numbers of tasks

#processor		greedy	ImpGrd	NoScore	Heuristic1	Heuristic2	Hybrid1	Hybrid2
4	FinTime	338824.25	56.25	0.25	0.25	0.25	0.25	0.25
	NumTask	2.00	0.00	18.00	0.00	0.00	0.00	0.00
6	FinTime	729230.97	406.17	0.17	0.17	0.17	0.17	0.17
	NumTask	3.60	0.00	9.20	0.00	0.00	0.00	0.00
8	FinTime	886931.98	704.55	0.27	0.27	0.27	0.27	0.27
	NumTask	4.57	0.29	6.57	0.29	0.00	0.00	0.00
10	FinTime	661935.66	337.43	0.10	0.10	0.10	0.10	0.10
	NumTask	3.33	0.44	3.11	0.00	0.00	0.00	0.00
12	FinTime	629248.27	1204.27	0.27	0.27	0.27	0.27	0.27
	NumTask	9.82	0.36	2.00	0.00	0.00	0.00	0.00
16	FinTime	826335.30	1819.16	0.23	0.23	0.23	0.23	0.23
	NumTask	4.80	0.53	1.20	0.27	0.27	0.53	0.27
20	FinTime	1145228.26	2262.68	0.47	1.52	0.47	0.26	2.37
	NumTask	1.79	0.32	1.16	0.11	0.11	0.21	0.00

Another way of evaluating schedules is to compare the variances of finishing times and variances of numbers of tasks. It reflects the ability of the algorithm to evenly distribute loads and tasks to processors, respectively. These values are shown in table 2. It can be observed that variants of *MOS* produced much smaller variances of finishing times than *Greedy* and *ImpGrd*. Meanwhile, *MOS* with heuristics produced much smaller variances of numbers of tasks than other algorithms. Generally, the smallest variances were produced by *Hybrid1*.

5.2 Results of a Single Particle Reconstruction Experiment

The *SADS* algorithm was tested in an actual single particle reconstruction experiment. The dataset had 2601 images of thermosome 5S alpha molecules, which were divided into 41 groups. *Hybrid1* was chosen as the heuristic. The program ran 10 rounds, with the rounding constant 0.01.

Table 3 gives the average relative errors between estimated and actual processing times for each round. Values for the first round are not listed because for the first round, the processing times were taken as the numbers of images. For other rounds, errors were small enough: all average relative errors were smaller than 1%. This indicates that the *SADS* algorithm estimates processing times accurately.

Table 3 Relative errors

Round	Relative Error (%)	Round	Relative Error (%)
1	-	6	0.764
2	0.855	7	0.702
3	0.865	8	0.708
4	0.993	9	0.772
5	0.801	10	0.812

Fig. 1 displays speedups of all rounds for 4 and 8 processors. For the first round, the speedups for all algorithms were relatively low because of the inaccurately estimated processing times. However, the speedups for *Hybrid1* were the highest in both cases. This was due to the effect of the secondary objective. Through other experiments, we found that the less accurate the estimation of processing times was, the more notable the advantages gained by the secondary objective were. *Hybrid1* and *NoScore* outperformed *Greedy* and *ImpGrd* in all cases, and they were also more stable than *Greedy* and *ImpGrd*. It can be noticed that *Hybrid1* was more stable than *NoScore*, and this was also because of the effect of the secondary objective.

Execution times for the scheduling scheme are shown in table 4. The longest processing time is no more than 0.1 seconds, which is trivial. Besides, execution times did not increase as the number of processors increased.

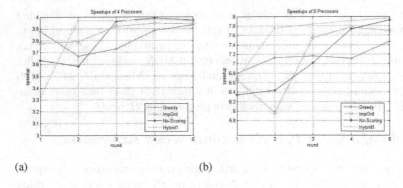

(a)   (b)

Fig. 1 Speedups of parallel classalign2

Table 4 Execution time

#PROCESSORS	TIME(S)	#PROCESSORS	TIME(S)
4	0.100	10	0.069
6	0.067	12	0.074
8	0.068	16	0.087

6 Discussion and Conclusions

In this study, a self-adaptive greedy scheduling scheme is introduced to parallelize one major component of EMAN. The core problem is a Multi-Objective Optimization on Identical Parallel Machines, whose primary objective is to minimize the makespan, and secondary objective to stabilize the schedule. Experimental results of a random dataset show that the scheduling scheme achieves the optimal primary and secondary objectives most of the cases, and the advantages of the scheme are more notable as the number of processors increases. Actual experiment of single particle reconstruction revealed that the scheduling scheme estimates processing times accurately. Better speedups can be obtained by the scheme, especially when the processing times are estimated less accurately. Finally, the time spent on the scheme is insignificant.

Besides EMAN, the scheduling scheme can also be applied to other similar programs. These programs should possess the following properties. First, they should be computing-intensive, and even a little load imbalance could result in significant extra execution time. Second, some methods can be adopted to estimate processing times of tasks, although such estimation may not be quite accurate. Third, the schedule is conducted in batch mode [16], on parallel computers with identical processors.

Acknowledgments. This work was supported by the National Natural Science Foundation for China Project under grant 90612019, 60752001, 60736012 and 60503060, and CAS knowledge innovation key project under grant KGGX1-YW-13.

References

[1] Frank, J.: Three-dimensional electron microscopy of macromolecular assemblies, pp. 1–13. Oxford University Press, Oxford (2006)

[2] Ludtke, S.J., Baldwin, P.R., Chiu, W.: EMAN: semiautomated software for high-resolution single-particle reconstructions. Journal of Structural Biology 128(1), 82–97 (1999)

[3] Frank, J., Radermacher, M., Penczek, P., Zhu, J., Li, Y., Ladjadj, M., Leith, A.: SPIDER and WEB: processing and visualization of images in 3D electron microscopy and related fields. J. Struct. Biol. 116, 190–199 (1996)

[4] Liang, Y., Ke, E.Y., Zhou, Z.H.: IMIRS: a high-resolution 3D reconstruction package integrated with a relational image database. J. Struct. Biol. 137, 292–304 (2002)

[5] Yu, X., Jin, L., Zhou, Z.H.: 3.88Å structure of cytoplasmic polyhedrosis virus by cryo-electron microscopy. Nature 452 (2008) doi:10.1038

[6] Scheres, S.H.W., Gao, H., Valle, M., Herman, G.T., Eggermont, P.P.B., Frank, J., mria Carazo, J.: Disentangling conformational states of macromolecules in 3D-EM through likelihood optimization. Nature Methods 4(1), 27–29 (2007)

[7] Yang, C., Penczek, P.A., Leith, A., Asturias, F.J., Ng, E.G., Glaeser, R.M., Frank, J.: The parallelization of SPIDER on distributed-memory computers using MPI. Journal of Structural Biology 157(1), 240–249 (2007)

[8] Hochbaum, D.S.: Approximation algorithms for NP Hard problems, pp. 1–17. PWS publishing company (1998)

[9] Graham, R.L.: Bounds for certain multiprocessing anomalies. Bell System Technical Journal 45, 1563–1581 (1966)

[10] Graham, R.L.: Bounds for multiprocessing timing anomalies. SIAM J. Appl. Math. 17, 416–426 (1969)

[11] Sahni, S.: Algorithms for scheduling independent tasks. J. Assoc. Comput. Mach. 23, 116–127 (1976)

[12] Hochbaum, D.S., Shmoys, D.B.: Using dual approximation algorithms for scheduling problems: practical and theoretical results. Journal of ACM 34(1), 144–162 (1987)

[13] Garey, M.R., Johnson, D.S.: Computers and Intractability: A guide to the theory of NP-Completeness. W. H. Freeman & Co., New York (1979)

[14] Lenstra, J.K., Shmoys, D.B., Tardos, E.: Approximation algorithms for scheduling unrelated parallel machines. Mathematical Programming 46, 259–271 (1990)

[15] Cormen, T.H., Leiserson, C.E., Rivest, R.L., Stein, C.: Introduction to algorithms, pp. 1043–1049. The MIT Press, Cambridge (2002)

[16] Maheswaran, M., Ali, S., Siegel, H.J., Hensgen, D., Freund, R.: Dynamic matching and scheduling of a class of independent tasks onto heterogeneous computing systems. In: 8th Heterogeneous Computing Workshop (HCW 1999) (April 1999)

Usage Distribution Coverage: What Percentage of Expected Use Has Been Executed in Software Testing?

Tomohiko Takagi, Kazuya Nishimachi, Masayuki Muragishi, Takashi Mitsuhashi, and Zengo Furukawa

Abstract. This paper shows that a UD (Usage Distribution) coverage criterion based on an operational profile measures the progress of software testing from users' viewpoint effectively. The UD coverage criterion focuses on not only the structure of software but the usage characteristics of users. Its measuring object is the software usage weighted with the probability of the operational profile, and intuitively it shows what percentage of expected use in operational environments has been tested. It solves the problem that usual coverage criteria focussing only on the structure of software are not substantiated by software reliability and do not identify the degree of an importance of each measuring object. This paper describes the definitions and examples of UD coverage.

1 Introduction

As software is becoming larger and more complicated, implementing high software reliability is the chief concern in the field of system development. The software reliability means the possibility that users will use software without encountering its failures in actual operational environments. One of the effective techniques for improving software reliability is software testing [1]. The software testing is the process to execute SUT (Software Under Test) in accordance with a specific predefined criterion and discover its failures. In order to measure the progress of software testing, coverage is widely used.

Tomohiko Takagi and Zengo Furukawa
Faculty of Engineering, Kagawa University
2217-20 Hayashi-cho, Takamatsu-shi, Kagawa 761-0396, Japan
e-mail: takagi@ismail.eng.kagawa-u.ac.jp, zengo@eng.kagawa-u.ac.jp

Kazuya Nishimachi, Masayuki Muragishi, and Takashi Mitsuhashi
JustSystems Corporation
BrainsPark Kawauchi-cho, Tokushima-shi, Tokushima 771-0189, Japan

R. Lee, N. Ishii (Eds.): Software Engineering, Artificial Intelligence, SCI 209, pp. 57–67.
springerlink.com   © Springer-Verlag Berlin Heidelberg 2009

Coverage derives from the ratio of objects that have been actually executed to all objects that should be executed in SUT. An object in this context is called a measuring object in this paper. For example, C0 used in structural testing is a well-known coverage criterion whose measuring objects are all the statements of SUT, and shows what percentage of them has been executed. High software quality cannot be implemented by using an only software testing technique, and likewise, true progress of software testing cannot be measured by using an only coverage criterion [6]; in other words, various viewpoints of coverage are required for measuring the progress. However, usual coverage criteria focus only on the structure of SUT as their measuring objects, then they are not substantiated by software reliability and do not identify the degree of an importance of each measuring object.

In this paper, we propose a UD (Usage Distribution) coverage criterion to solve this problem. Its measuring object is the usage of SUT weighted with the use probability of actual users, and intuitively UD coverage shows what percentage of expected use in operational environments has been executed in software testing. Section 2 describes the problem of usual coverage criteria, and then Sect. 3 gives the definitions and examples of UD coverage. Section 4 describes consideration, and finally Sect. 5 shows related work.

2   Problem of Usual Coverage Criteria

Many coverage criteria have been proposed and also used in the field of software development (examples are shown in Sect. 5). The definition of usual coverage is as follows.

Definition 1. Usual coverage

$$C(O') = \frac{|O'|}{|O|},$$
(1)

where O is a set of measuring objects, and O' is a set of elements of O that have been tested; $|O|$ represents the number of elements of O.

Usual coverage criteria focus only on the structure of SUT such as source codes and models in specifications when preparing their measuring objects. Therefore there are the following two problems:

- They are not substantiated by software reliability; i.e., they show the quality of software development work, not software reliability. Even if coverage reaches 100%, it does not mean that SUT is on a specific level of software reliability. They lack the viewpoint of users, and therefore it is difficult to know when to ship SUT.
- They do not identify the degree of an importance of each measuring object. In Def. 1, the numerator is increased by one when an unexecuted measuring object is executed. This means that a measuring object that relates to a key facility and one that relates to an optional facility are treated equally. However, testing the former rather than testing the latter can contribute to the progress.

Solving these problems requires a new coverage criterion that focuses on not only the structure of SUT but the usage characteristics of users.

3 UD Coverage Criterion

We propose a UD coverage criterion in order to solve the problems described in Sect. 2. Its measuring object is the usage of SUT weighted with the use probability of actual users, and intuitively UD coverage shows what percentage of UD (i.e., expected use in operational environments) has been executed in software testing. It is calculated based on a probabilistic state machine diagram called an operational profile [7] or a usage model [12, 13]. Our method is applicable to various functional testing techniques on levels of integration testing, system testing, acceptance testing and regression testing that can be managed on the state machine diagram; they do not need to be based on an operational profile.

Construction of a state machine diagram, survey of operational environments, and development of an operational profile are required before calculation of UD coverage. The operational profile is originally used as a model for generating test cases, then our method derives the operational profile suitable for measuring UD coverage. In addition, the formal definition of UD coverage is related closely to the operational profile. Therefore, the following subsections provide the definitions and simple examples for all the steps.

3.1 Construction of a State Machine Diagram

The operational procedure of SUT is defined as a state machine diagram [8]. In the state machine diagram, a state represents a stage of use, an event represents a user operation, and a transition means that the event changes the state. It is constructed from the viewpoint of users, and does not require the detailed description of internal states of SUT. Abnormal conditions such as operation mistakes are defined if their UD coverage has to be calculated. In a requirement definition process, the state machine diagram can be constructed as a small part of the specifications of SUT, or in a test process, it can be constructed based on some specifications with the (additional) purpose of reviewing them.

Below is the formal definition of a state machine.

Definition 2. State machine

$$SM = \{S, i, F, E, T\}, \tag{2}$$

where S is a set of states in SM; i is an initial pseudo state in SM; F is a set of final pseudo states in SM; E is a set of events in SM; T is a set of transitions in SM; an element of T represents a unique list <*from-state, event, to-state*> and satisfies $<x, y, z> \in T \rightarrow x \in \{S, i\} \land y \in E \land z \in \{S, F\} \land \neg(x = i \land z \in F)$.

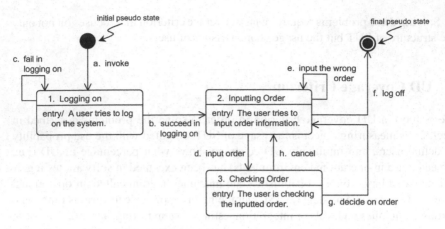

Fig. 1 State machine diagram of an internet shopping system

A simple example of a state machine diagram is shown in Fig. 1. This internet shopping system is assumed to be used in the following way:

Initial pseudo state	A user of the internet shopping system invokes a Web browser.
State 1	The user tries to log on the system. If the user succeeds in logging on, he can input his order information. If not, he retries to log on.
State 2	When the user finishes inputting his order information, the system checks it automatically. If the system finds a wrong input, the user can retry to input. When the user finishes shopping, he logs off the system.
State 3	The user checks the inputted order. If there are no problems about the order, he decides on it. If not, the user cancels the order and can try to input another order.

3.2 Survey of Operational Environments

Expected operational environments of SUT are surveyed to gather field data that are useful for calculating probability distributions in the state machine diagram. Examples of the field data are complete operating logs for earlier versions of SUT, raw data that result from business, and so on. They are then converted into a form that can be mapped to the state machine diagram. The survey of operational environments would be expensive, and therefore it is preferable that the field data should be converted into a convenient form to reuse for other similar SUT. There is a method of using constraints for mathematical programming [10], but our method uses transition sequences as such a form.

Below are the formal definitions of transition sequences TS and of field data FD.

Fig. 2 Field data of an
internet shopping system

●a1b2d3g2f●
●a1c1b2d3h2d3g2f●
●a1b2e2e2d3g2e2d3g2f●
●a1b2d3g2d3g2d3g2f●
●a1c1c1b2d3g2d3h2d3g2f●
●a1b2f●
●a1b2d3g2d3g2d3g2d3g2d3g2f●
●a1c1b2d3g2f●

Definition 3. Transition sequences in *SM*

$$TS = \{< t_1, t_2, \cdots, t_n >| \forall j (1 \le j \le n \to t_j \in T)$$
$$\wedge t_1[1] = i \wedge t_n[3] \in F$$
$$\wedge \forall k (1 \le k \le n-1 \to t_k[3] = t_{k+1}[1])\},\qquad(3)$$

where $x[y]$ represents the yth element of list x (e.g., $< x, y, z > [2] = y$); in this paper
we use special mathematical symbols of the literature [4]. TS is defined as a set of
all the sequences that are consecutive transitions from the initial pseudo state to the
final pseudo state in *SM*.

Definition 4. Field data that were converted into transition sequences in *SM*

$$FD = \{< ts_1, ts_2, \cdots, ts_n >| \forall j (1 \le j \le n \to ts_j \in TS)\}.\qquad(4)$$

Figure 2 is the example of the field data expressed as the transition sequences in the
state machine diagram of the internet shopping system (Fig. 1). A state and event
are represented by a numeric and alphabet character respectively, and the description of a transition sequence is simplified for convenience, e.g., 1b2d3g2 instead of
<1b2><2d3><3g2>.

3.3 Development of an Operational Profile

Probability distributions in the state machine diagram are calculated based on the
gathered field data in order to complete an operational profile of SUT. When the
operational profile is accurate (i.e., when the operational profile is close to behavior
of actual users), accurate evaluation of UD coverage can be obtained. There is the
method of developing an accurate operational profile based on a high-order Markov
chain [11], and it can be incorporated into this step when the transition sequences
are gathered as the field data.

Below is the formal definition of an operational profile *OP*.

Definition 5. Operational profile that is developed based on a state machine *SM* and
an element fd of field data *FD*

$$OP = \{S, i, F, E, T_{op}\},\qquad(5)$$

Fig. 3 Operational profile of
an internet shopping system

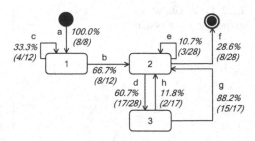

where T_{op} is a set of probabilistic transitions in OP; an element of T_{op} represents
a unique list *<from-state, event, transition probability, to-state>* and satisfies the
following:

$$< x,y,p,z >\in T_{op} \rightarrow < x,y,z >\in T \wedge p = \frac{\gamma/fd \downarrow \{t \mid t =< x,y,z >\}}{\gamma/fd \downarrow \{t \mid t[1] = x \wedge t \in T\}} , \quad (6)$$

where γ/x represents flattened x, and $x \downarrow y$ represents the number of y in x (e.g.,
$\gamma/ << x,y >,< z >>=< x,y,z >$, $< x,y,x > \downarrow x = 2$).

Figure 3 shows the operational profile developed from the state machine diagram
(Fig. 1) and transition sequences (Fig. 2) of the internet shopping system.

3.4 Calculation of UD Coverage

Each test case executed in arbitrary testing techniques is converted into a transition
sequence that starts with the initial pseudo state and terminates with the final pseudo
state in the operational profile, and then a total product of a series of its transition
probabilities is calculated. The total product is called a weight of each test case
in this paper. UD coverage is a total sum of the weights in a set of the transition
sequences (i.e., the test cases) of which a test history consists.

Below are the formal definitions of test cases TC, of a weight $w(tc)$, and of UD
coverage $UC(TH)$.

Definition 6. Test cases that were converted into transition sequences in OP

$$TC = \{< t_1, t_2, \cdots, t_n >\mid \forall j(1 \leq j \leq n \rightarrow t_j \in T_{op})$$
$$\wedge t_1[1] = i \wedge t_n[4] \in F$$
$$\wedge \forall k(1 \leq k \leq n-1 \rightarrow t_k[4] = t_{k+1}[1])\} , \quad (7)$$

TC is defined as a set of all the sequences that are consecutive transitions from the
initial pseudo state to the final pseudo state in OP.

Definition 7. A weight of an element tc of TC

$$w(tc) = \prod_{k=1}^{\#tc} \gamma/tc[4k-1] , \quad (8)$$

where $\sharp x$ represents the length of list x (e.g., $\sharp < x, y, z >= 3$). $w(tc)$ is defined as a total product of transition probabilities in tc.

Definition 8. UD coverage of a test history TH

$$UC(TH) = \sum_{k=1}^{N} w(th_k) , \qquad (9)$$

where TH satisfies $TH \subseteq TC$, and is defined as $TH = \{th_1, th_2, \cdots, th_N\}$. $UC(TH)$ is a total sum of weights of elements in TH.

The UD coverage shows the extent of a test history on expected use of SUT. If a facility of high use probability is tested, the UD coverage increases greatly, but if a facility that is not used at all is tested, it does not increase at all. It is usually impossible that UD coverage reaches 100%, because almost all state machine diagrams have some loops and therefore TC would have infinite elements.

In an actual test process, software engineers usually try to assure the quality of SUT by using a subset of TC. In this case, the engineers can define the subset for each test process, and evaluate UD coverage on the subset.

Below is an example of the definition of a subset.

Definition 9. A subset of TC on condition that the identical transition is not executed more than 10 times in each test case

$$TC' = \{tc' \mid tc' \in TC \land \forall k (2 \leq k \leq \sharp tc' - 10 \rightarrow tc' \downarrow tc'[k] \leq 10)\} . \qquad (10)$$

When $TC - TC' \neq \{\}$ is true, UD coverage is defined as the following.

Definition 10. UD coverage of a test history TH'

$$UC'(TH') = \frac{\sum_{k=1}^{N} w(th'_k)}{\sum_{j=1}^{M} w(tc'_j)} , \qquad (11)$$

where TH' satisfies $TH' \subseteq TC'$, and they are expressed as $TH' = \{th'_1, th'_2, \cdots, th'_N\}$, $TC' = \{tc'_1, tc'_2, \cdots, tc'_M\}$. $UC'(TH')$ is given when dividing a total sum of weights of elements in TH' by a total sum of weights of elements in TC'.

Definition 10 is the generalized definition of UD coverage, since it is transformed to Def. 8 when $TH' = TH$ and $TC' = TC$. UD coverage has the advantage of the flexible definitions that can be extended according to the purpose and situation of testing, as shown in Def. 9 and 10.

The following is the simple example of UD coverage calculation. It is assumed that the internet shopping system has been executed by three test cases shown in Fig. 4. The test cases are expressed as the transition sequences of the state machine diagram shown in Fig. 1, then it is straightforward to convert them into the transition sequences of the operational profile shown in Fig. 3. The weights of test case No.1, No.2 and No.3 are given by the following calculations:

Fig. 4 Test history of an
internet shopping system

(No.1)	●a1b2d3g2f●
(No.2)	●a1c1b2d3h2f●
(No.3)	●a1b2d3g2f●

(No.1) $1.000 \times 0.667 \times 0.607 \times 0.882 \times 0.286 \approx 0.102$
(No.2) $1.000 \times 0.333 \times 0.667 \times 0.607 \times 0.118 \times 0.286 \approx 0.005$
(No.3)   This is same as No.1.

In the above, the test case No.3 has been already executed as No.1, and consequently the UD coverage is about 10.7% (10.2% plus 0.5%). Thus if no failures
are discovered by using these test cases, it means that 10.7% of the expected use
in operational environments has been assured by them, and the software reliability
per use (i.e., per transition sequence in the operational profile) is estimated to be not
less than 10.7%. In addition, the test case of a great weight such as No.1 covers the
expected use more widely than other test cases, and it contributes to the progress of
software testing greatly.

4   Consideration

As shown below, the problems described in Sect. 2 have been solved by UD coverage:

- The UD coverage is substantiated by software reliability. For instance, Sect. 3
  has shown that the software reliability per use is estimated to be not less than
  10.7% since the UD coverage is 10.7%. This estimation is true, because the test
  cases that cover 10.7% of expected use are executed and can ensure that no failures occur in the covered use. The UD coverage has the viewpoint of users, and
  therefore it helps software engineers to know when to ship SUT.
- The UD coverage identifies the degree of an importance of each measuring object. For instance, Sect. 3 has shown that UD coverage increases greatly if the test
  case of a great weight is executed. The test case of a great weight corresponds
  to the usage that is frequently executed in operational environments. The failure
  that relates to such usage is a serious risk in the sense that it occurs frequently.
  When a testing technique to increase the UD coverage efficiently is selected, the
  risk is reduced at the early stage.

In this study, we have developed HOUMA (High-Order Usage Model Analyzer)
system that automates the development of an operational profile, the generation of
test cases, the calculation of UD coverage and so on. The overview of HOUMA is
shown in Fig. 5.

Figure 6 shows the growth of UD coverage when random testing technique (using
uniform distribution) and statistical testing technique (using the operational profile)
[12, 13] are applied to the example of Sect. 3. The statistical testing increases UD
coverage more rapidly than the random testing, and therefore the statistical testing is preferable to the random testing. In the statistical testing, the higher the use

Fig. 5 Overview of HOUMA

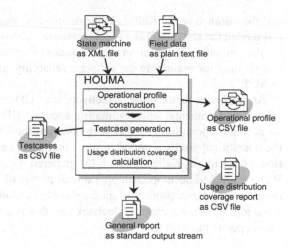

Fig. 6 The growth of UD coverage

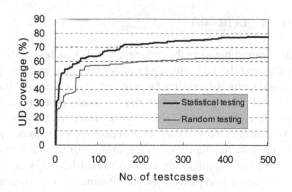

probability of a facility is, the earlier the facility is tested (i.e., the greater the weight of a test case is, the earlier the test case is executed), and Fig. 6 demonstrates it well.

5 Related Work

This section describes related work.

The UD coverage criterion is based on a state machine. An example of usual coverage criteria based on a state machine is N-switch [2]. It is the measure for covering transition sequences of length $N + 1$ in a state machine. Offutt et al. [9] proposed four coverage levels for a state machine; i.e., transition coverage level, full predicate coverage level, transition-pair coverage level, and complete sequence level. The difference between UD coverage and usual coverage is that the former has the viewpoint of software reliability. In regard to the criteria of the software reliability, the literature on statistical testing technique [13] reported the expected number of transitions between failures, the probability of discovering a failure per test case, and the distance of the Kullback discriminant between an operational profile and a test model; the test model is a special operational profile developed by mapping test results (including information about failures) to a state machine diagram. They are criteria only for evaluating the software reliability, and not for evaluating coverage of SUT.

An important factor in the effectiveness of UD coverage is how to survey operational environments, since estimating accurate UD coverage requires developing an accurate operational profile. The concrete methods of gathering field data for the operational profile reported in some literature, and they provide useful information. For example, Kallepalli et al. [5] gathered data by using logging systems on a Web server, and developed an operational profile of a Web application. In order to develop operational profiles of medical systems, Hartmann et al. [3] gathered operation logs by using a capture/playback tool that is a sort of test tool when expected users operate them on trial.

6 Conclusion

In order to measure the progress of software testing from users' viewpoint, we have proposed the UD coverage that shows what percentage of expected use in operational environments has been tested. It is calculated based on an operational profile of SUT. The operational profile is a model that represents UD as a probabilistic state machine diagram, and if it is not accurate, its resultant UD coverage is not accurate. The strong points of the UD coverage are to estimate the software reliability of SUT, to identify the degree of an importance of each test case.

We have applied this method to real-life software, and its result shows that UD coverage would provide useful information for software engineers. Future work includes applying this method to software development several times and analyzing collected data statistically in order to find good enough coverage. Also, we plan to improve existing test case generation algorithm to increase UD coverage efficiently.

References

[1] Beizer, B.: Software Testing Techniques, 2nd edn. Van Nostrand Reinhold (1990)
[2] Chow, T.S.: Testing software design modeled by finite-state machines. IEEE Transactions on Software Engineering SE-4(3), 178–187 (1978)
[3] Hartmann, H., Bokkerink, J., Ronteltap, V.: How to reduce your test process with 30% – the application of operational profiles at philips medical systems. In: Supplementary Proceedings of 17th International Symposium on Software Reliability Engineering, CD-ROM (2006)
[4] Hoare, C.A.R.: Communicating Sequential Processes. Prentice-Hall, Englewood Cliffs (1985)

[5] Kallepalli, C., Tian, J.: Usage measurement for statistical web testing and reliability analysis. In: Proceedings of 7th International Software Metrics Symposium, London, pp. 148–158 (2001)

[6] Kaner, C., Bach, J., Pettichord, B.: Lessons Learned in Software Testing. John Wiley & Sons, Chichester (2002)

[7] Musa, J.D.: The operational profile. Computer and system sciences, NATO ASI Series F 154, 333–344 (1996)

[8] Object Management Group, Unified Modeling Language (2007), http://www.uml.org/

[9] Offutt, J., Abdurazik, A.: Generating tests from uml specifications. In: Proceedings of 2nd International Conference on the Unified Modeling Language, pp. 416–429 (1999)

[10] Poore, J.H., Walton, G.H., Whittaker, J.A.: A constraint-based approach to the representation of software usage models. Information and Software Technology 42(12), 825–833 (2000)

[11] Takagi, T., Furukawa, Z.: Construction method of a high-order markov chain usage model. In: Proceedings of 14th Asia-Pacific Software Engineering Conference, pp. 120–126 (2007)

[12] Walton, G.H., Poore, J.H., Trammell, C.J.: Statistical testing of software based on a usage model. Software Practice and Experience 25(1), 97–108 (1995)

[13] Whittaker, J.A., Thomason, M.G.: A markov chain model for statistical software testing. IEEE Transactions on Software Engineering 20(10), 812–824 (1994)

Concept Based Pseudo Relevance Feedback in Biomedical Field

Vahid Jalali and Mohammad Reza Matash Borujerdi

Abstract. Semantic information retrieval is based on calculating similarity between concepts in a query and documents of a corpus. In this regard, similarity between concept pairs is determined by using an ontology or a meta-thesaurus. Although semantic similarities often convey reasonable meaning, there are cases where calculated semantic similarity fails to detect semantic relevancy between concepts. This problem often occurs when concepts in a corpus have statistical dependencies while they are not conceptually similar. In this paper a concept-based pseudo relevance feedback approach is introduced for discovering statistical relations between concepts in an ontology. Proposed approach consists of adding extra concepts to each query based on the concepts in top ranked documents which are rendered as relevant to that query. Results show that using conceptual relevance feedback would increase mean average precision by 19 percent compared to keyword based approaches.

Keywords: Concept-based Pseudo Relevance Feedback, Information Retrieval, Medical Subject Headings, Semantic Similarity Measures.

1 Introduction

The main idea of semantic information retrieval is to overcome shortcomings of keyword based approaches. There are cases where two documents talk about the same subject with different vocabularies. In these situations systems which rely totally on keyword-based similarities, will judge two documents as non-relevant. However in general the lack of similar terms does not indicate the lack of semantic similarity. Each semantic information retrieval system exploits a kind of similarity measure for calculating conceptual relevancies in a corpus. There are three major approaches named edge-counting measures, information content measures and feature-based measures for calculating semantic similarity using an ontology.

Edge-counting approach which is used in proposed semantic information retrieval algorithm, calculates similarity between concepts based on their position in

Vahid Jalali and Mohammad Reza Matash Borujerdi
Advanced Artificial Intelligence Laboratory, Computer Engineering Department,
Amirkabir University of Technology, Tehran, Iran
e-mail: {vjalali, borujerm}@aut.ac.ir

R. Lee, N. Ishii (Eds.): Software Engineering, Artificial Intelligence, SCI 209, pp. 69–79.
springerlink.com   © Springer-Verlag Berlin Heidelberg 2009

a hierarchy. Although edge-counting based approaches capture semantic similarities which can be inferred from the structure of used ontology, they cannot detect semantic relevancies between those concept pairs which are not located near each other in the ontology. On one hand this inconsistency may occur because of deficiency or incompleteness of used ontology, and on the other hand it may be faced just because of inherent characteristics of the corpus which is used by the system.

For example consider "Neuroleptic Malignant Syndrome" and "Antipsychotic Agents" which are two concepts in MeSH (Medical Subject Headings) meta-thesaurus. In this case, although former concept is a syndrome associated primarily with the use of latter one, there is no explicit relation between these two concepts in MeSH tree structure. Hence any approach considering merely MeSH hierarchy fails to detect their relevancy. As another example consider concepts "Anorexia Nervosa" and "Bulimia" which are two kinds of diseases, one related to lack of appetite and the other related to eating an excess amount of food in a short period of time. In a specific corpus it is seen that the occurrence of one of these concepts in a document, is often followed by the occurrence of the other. Here again two concepts are related to each other despite the fact that they are not semantically relevant based on the hierarchical structure of used ontology.

These observations give the insight that using statistical relations between different concepts, in addition to their semantic relatedness would be beneficial in information retrieval. The mechanism used in this experiment for detecting relevant concepts for a query is based on pseudo relevance feedback which is a well known technique in information retrieval. In general three types of feedback can be distinguished including: explicit feedback, implicit feedback and pseudo feedback. While explicit and implicit feedbacks depend on user's choices or behaviors, pseudo relevance feedback is performed without the interaction of users. Because gathering lots of users and logging their interactions with system is not possible in most cases; hence proposed approach in this paper is based on pseudo relevance feedback for detecting statistical dependencies between concepts.

The paper is organized as follows. Section 2 reviews related work in field of semantic information retrieval and pseudo relevance feedback for medical field. Section 3 describes proposed approach for performing semantic retrieval and the role of pseudo relevance feedback in it. Experiment and results are reported in Section 4. Section 5 raises extra discussions or clarifications which are not explained through the paper. Section 6 concludes the paper and introduces extra directions for future work.

2   Related Work

Although much work is done in semantic information retrieval and query expansion, the idea of using concept-based pseudo relevance feedback as it is introduced in this paper to the best of our knowledge is an innovative idea in this field.

Query expansion techniques can broadly be classified into three categories:

1. Collection based or global analysis: use context global of terms in collection to find out similar terms with query terms [4].
2. Query based or local analysis: the context of terms is reduced to smaller subsets of information which is given from relevance feedback or pseudo relevance feedback [2] and collaboration information like user profile, passed queries [3].
3. Knowledge based approach: is the exploration of the knowledge in external knowledge sources, mostly with general domain thesaurus like Wordnet [14, 10]. They explore semantic links in this thesaurus in order to find out in the related terms of query concepts to expand, but results aren't positive somehow due to word sense ambiguities.

As a matter of fact, proposed approach in this paper belongs to second group of query expansion techniques. Proposed method differs from other pseudo relevance feedback approaches considering the fact that it treats additive terms to initial query as new concepts instead of new keywords. Hence, newly added terms are merely used for re-calculating concept-based scores and not keyword-based ones. In continue most eminent works on query expansion and semantic information retrieval in biomedical field are reviewed.

Hersh [5] found that unconstrained UMLS based synonym and hierarchical query expansion on the OHSUMED collection degraded aggregate retrieval performance, but some specific instances of Query Expansion improved individual query performance. He noted that there was a role for investigation of specific cases where Query Expansion is successful. Overall, there are a wide variety of possible relationships usable for expansion. Basic categories include synonymy, hierarchy, and associative type relationships.

Aronson [1] examined both identification of UMLS Meta-thesaurus in user's query and using them in query expansion and retrieval feedback. He advocated using a combination of these approaches in order to get best results.

Srinivasan [13] tested expansion using three alternatives: an inter-field statistical thesaurus, expansion via retrieval feedback and expansion using a combined approach. The study compared retrieval effectiveness using the original unexpanded and the alternative expanded user queries on a collection of 75 queries and 2334 MEDLINE citations. Retrieval effectiveness was assessed using eleven point average precision scores (11-AvgP) and She concluded that the combination of expansion using the thesaurus followed by retrieval feedback gives the best improvement of 17% over a baseline performance of 0.5169 11-AvgP. The difference between relevance feedback used in Srinivasan's approach and what is introduced in this paper is that Srinivasan's relevance feedback is merely keyword based and instead of adding new concepts to initial queries extra keywords are added to them.

Mao [11] introduced a phrase-based vector space model for information retrieval in biomedical field. in his approach, in addition to syntactic similarities between documents and queries, their conceptual similarities were also taken into account. Conceptual similarity in Mao's approach was calculated based on concepts distances in a hypernym hierarchy. Mao reported 16% improvements in retrieval accuracy compared to stem-based model in OHSUMED test collection. Proposed approach in this paper is similar to that of Mao's in considering conceptual relatedness of queries and documents in addition to their keyword-based similarities. However Mao did not pay attention to co-occurrence of concepts and their statistical dependencies which would be another source of relevancy detection in semantic information retrieval.

Yoo and Choi [15] proposed automatic query expansion with pseudo relevance feedback using OHSUMED test collection. They tested several term sorting methods for the selection of expansion terms and proposed a new term re-weighting method for further performance improvements. Through the multiple sets of test, they suggested that local context analysis is probably the most effective method of selecting good expansion terms from a set of MEDLINE documents given enough feedback documents.

Houston [6] compared 3 different thesauri for medical query expansion, consisting of corpus terms, MESH terms, and UMLS terms. He found no significant difference between them, but also little overlap, supporting a case for further exploration to discover the significant features of each term set, and use this information to combine their strengths.

Leroy [8] found that UMLS relationships provide better QE than that of Word-Net in the medical field. Leroy Earlier query expansion works focused on a technique called Deep Semantic Parsing. This technique combined co-occurrence and UMLS semantic relationship information, and derives from the idea that a given semantic relationship does not apply to every combination of members of two groups connected by this relationship. This technique used these found semantic network relationships to limit co-occurrence relationships used for expansion. The system was evaluated over two types of query, sourced from medical librarians and cancer specialists. Cancer specialists had smaller queries; they were more precise, and had lower recall. Expansion improved recall for more comprehensive queries, while precision dropped more for smaller, more precise queries.

3   Proposed Approach

Pseudo relevance feedback used in this experiment is different from other kinds of relevance feedback, considering the fact that instead of terms, concepts are added to initial query. These additive concepts are then used to return semantically similar documents to user's query by using an edge-counting based similarity measure. In proposed approach additional concepts are extracted by examining concepts frequencies in top ranked documents returned by a hybrid information retrieval system for each query. The hybrid retriever is based on combining keyword based and semantic scores for query-document pairs, where keyword based scores are calculated by Lucene search engine and semantic scores are calculated using MeSH tree structure which is proved to be beneficial in improving retrieval performance [12], and Li similarity measure [9]. Concept extraction algorithm used in this experiment is the same as one used in [7] which is based on using Gate natural language processing framework and MeSH ontology together.

3.1 Keyword-Based Scoring Method

Keyword based scoring algorithm is based on a combination of vector space and Boolean model for information retrieval by using Lucene search engine. Each document in this experiment has a title and abstract which are extracted from Medline publications. Indexing of documents is performed by applying a boosting factor to title terms which is 5 times greater than that of terms in abstract part. It should be mentioned that no other extra terms or even stemming algorithms are used for document indexing.

3.2 Semantic Retrieval Algorithm

Semantic similarity measure used in proposed method is based on Li's measure [9], which combines the shortest path length between two concepts C_1 and C_2, L, and the depth in the taxonomy of the most specific common concept C, H, in a non-linear function. In equation (1) Li similarity measure is exhibited.

$$Sim_{Li}(C_1, C_2) = e^{-\alpha L} \cdot \frac{e^{\beta H} - e^{-\beta H}}{e^{\beta H} + e^{-\beta H}} \tag{1}$$

where $\alpha \geq 0$ and $\beta > 0$ are parameters scaling the contribution of shortest path length and depth respectively. Based on Li's work, the optimal parameters are $\alpha = 0.2$ and $\beta = 0.6$. This measure is motivated by the fact that information sources are infinite to some extend while human compares word similarity with a finite interval between completely similar and totally non-relevant. Intuitively the transformation between an infinite interval to a finite one is non-linear. It is thus obvious that this measure scores between 1 (for similar concepts) and 0.

Concepts of queries in proposed system are extracted using natural language processing techniques and MeSH ontology, while concepts of documents are previously assigned by human experts. Each concept in MeSH tree structure can have different locations. For calculating semantic similarity between a query and a set of documents all of these position variations are taken into account and each concept from query is examined against those of each document. Having calculated different scores for each query concept, the maximum score is assigned to that concept in related query-document pair.

For each query-document pair in the corpus, the maximum score calculated in previous step and the average of these scores are stored. These two factors later combine with each other to form the semantic score of proposed retrieval algorithm. In equation (2) the combination of these factors is demonstrated.

$$SemanticSim(q_i, d_j) = \frac{\sum_{k=1}^{m} \max_{l=1,\dots,n} \max(Sim_{Li}(qc_k, dc_l))}{m} + floor(\frac{\max_{k=1,\dots,m} \max_{l=1,\dots,n} (Sim_{Li}(qc_k, dc_l))}{0.9}) \tag{2}$$

Where q_i is i-th query, d_j is j-th document, m and n are the number of concepts in i-th query and j-th document respectively, and also qc_k is m-th concept in i-th query and dc_l is l-th concept in j-th document.

The reason for using second factor in equation (2) is that queries in selected test collection are often small and contain only one concept. In this case it is possible that average score of a query-document pair with two query concepts is assigned a lower score than that of a query-document pair with one query concept just because of low score assigned to one of its two concepts. Furthermore it is observed that query-document pairs which are actually relevant to each other in most cases share identical concepts, hence there is a great possibility that those pairs with score in range of 0.9 to 1 are among those pairs judged as relevant. Experiment shows that when the number of query concepts increases second factor can be eliminated without causing noticeable change in performance of the system.

In this experiment, for calculating semantic relatedness between query-document pairs a technique similar to classic stop word elimination is also used. In this regard semantic scores are calculated only for those concepts which their document frequency is lower than a specific threshold. The threshold used for stop concept elimination is 10000 for OHSUMED collection. This means that concepts which are repeated in more than 10000 documents would not be of great importance in determining relevancies of queries and documents and hence they are not interfered in calculating semantic relatedness of query and documents.

Final scores of first retrieval phase are calculated by using a symmetric average between semantic and keyword based scores which is shown in equation (3). Having calculated final scores, based on the number of documents returned for a specific query, a percent of top ranked documents are selected and their concepts frequencies are examined in order to extract most relevant concepts and add them to the query.

$$Sim(q_i, d_j) = \frac{SemanticSim(q_i, d_j) + LucenceSim(q_i, d_j)}{2} \qquad (3)$$

3.3 Applying Pseudo Relevance Feedback

At this stage two groups of concepts would not be selected as additive concepts. The first group contains concepts with document frequency greater than 10000, and the other includes concepts which there exists a relating concept in initial query for them so that their semantic similarity is greater than a specific threshold. This threshold in proposed approach is 0.7 for Li similarity measure between two concepts. The reason for this selection is the intuition that adding concepts with greater semantic similarity to the query has no effect on overall precision of the system, because there is already a concept in initial query which its location in MeSH hierarchy is close to the location of this candidate concept.

Once extracted concepts are added to initial query, semantic and final scores of query-document pairs are re-calculated like what is explained previously in this section. As it is mentioned before in calculating final scores after injecting additive concepts to initial queries second factor of equation (2) would be eliminated.

4 Experiment and Results

Proposed approach in this paper is tested on OHSUMED test collection. OHSUMED is a large test collection for information retrieval systems which contains about 348000 documents, 106 queries and a set of relevance judgments.

The reference collection is a subset of the MEDLINE database. Each reference contains a title, an optional abstract, a set of MeSH headings, author information, publication type, source, a MEDLINE identifier, and a sequence identifier. 14,430 references out of the 348K are judged by human experts to be not relevant, possibly relevant, or definitely relevant to each query.

The knowledge source used in this experiment for calculating semantic similarity between concepts is MeSH meta-thesaurus. MeSH is a huge controlled vocabulary for the purpose of indexing journal articles and books in the life sciences, Created and updated by the United States National Library of Medicine (NLM). This vocabulary at the time contains 24767 subject headings, also known as descriptors. Most of these descriptors are accompanied by a short definition, links to related descriptors, and a list of synonyms or very similar terms (known as entry terms)

In first part of the experiment 101 queries are used for retrieving relevant documents. In addition, those documents which have empty abstracts are eliminated from judgments and a number of 10718 query-document pairs (from 14,430 judged pairs) are analyzed for retrieval task. For each of these 10718 pairs both

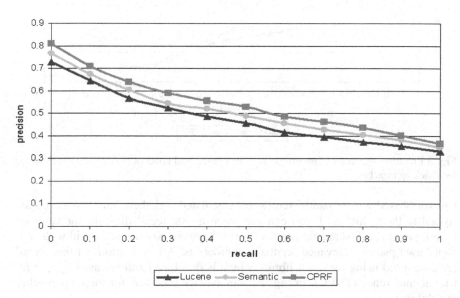

Fig. 1 Precision vs. recall diagram for keyword based, semantic and pseudo relevance feedback approaches

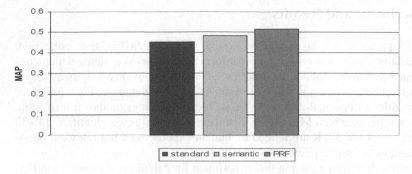

Fig. 2 Mean average precision for keyword based, semantic and concept based pseudo relevance feedback approaches

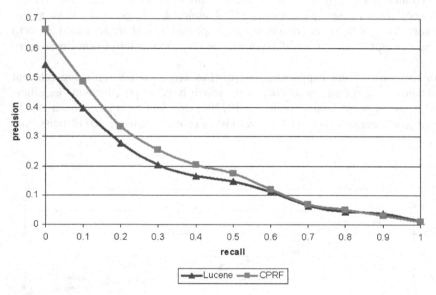

Fig. 3 Precision vs. recall diagram for keyword based and concept based pseudo relevance feedback approaches

keyword-based and semantic scores are calculated and then final scores are assigned to them. In Fig. 1 you can see precision vs. recall diagram for keyword based, pure hybrid (semantic combined with keyword based) and hybrid with concept based pseudo relevance feedback approaches. All evaluations in this regard are calculated using trec-eval library which is developed and released by text retrieval conference (TREC). In Fig. 2 mean average precision for three approaches is exhibited.

In second part of experiment the whole corpus is searched for finding relevant documents. First results are those returned by Lucene with scores greater than 0.1 assigned to them. In this regard 10357 documents are rendered relevant for 95 queries from OHSUMED test collection. Total number of relevant documents over all queries in this case would be 2993, from which 1135 documents are retrieved in result set. Having all rendered pairs by Lucene, semantic scores for all these pairs are calculated like what is explained previously in this section and extra concepts are added to initial queries by concept-based pseudo relevance feedback introduced in this paper. Semantic scores are re-calculated by using additive concepts from relevance feedback and total scores are calculated according to equation (3). Results of second part of experiment are exhibited in Fig. 3 in form of precision vs. recall diagrams for keyword based and concept based pseudo relevance feedback approaches.

Difference between first part and second part of the experiment is that, in first part retrieval process is performed on a set containing all relevant pairs and a number of non-relevant pairs, while in second part relevant pairs should be discovered from the whole corpus. Results show 19% improvements in term of mean average precision compared to keyword-based approach when the whole corpus is examined for finding relevant documents.

5 Discussion

Query expansion techniques which exploit external knowledge sources such as ontologies in expansion process are very similar to those semantic retrieval algorithms which use the same knowledge source for calculating edge-counting based semantic similarity measures. In this regard, using enough expansion levels and applying appropriate boosting factors to additive terms from each level would simulate a kind of semantic similarity measure between queries and documents. This is why we have studied query expansion methods and semantic information retrieval approaches together in section 2.

Assigning a boosting factor to title terms of documents, which is 5 times greater than that of abstract part, is determined by examining different boosting factors in retrieving relevant documents with IDs ranging from 330K to 348K in OHSUMED test collection.

Threshold for stop concept elimination which is equal to 10000 in this experiment is determined by a human expert. Concepts are sorted descending by their corresponding frequency over the whole corpus and presented to a human expert to determine a threshold for identifying discriminative or indiscriminative concepts.

In OHSUMED test collection queries are composed of two parts, which are information request and patient information respectively. In this experiment patient information part of each query is ignored and only information request part is used in retrieval task. There are many cases in which additive concepts extracted from local analysis of top ranked retrieved documents are identical to concepts of patient information part of each query.

There are also a noticeable amount of additive concepts which can be detected by using MeSH relations other than those of hypernym hierarchy. As an example consider "Neuroleptic Malignant Syndrome" and "Antipsychotic Agents" concepts which are introduced in first section of this paper. If one follows scope note attribute of "Neuroleptic Malignant Syndrome" concept in MeSH ontology he would face "Antipsychotic Agents" as a related concept. This observation gives the insight for working on different relations and attributes of MeSH ontology in addition to co-occurrence of concepts in local analysis of retrieved documents.

6 Conclusion and Future Work

In this paper a concept based pseudo relevance feedback algorithm for information retrieval in biomedical field is introduced. Proposed approach is based on ranking results by a combination of keyword based and semantic methods and use top ranked documents to extract additive concepts for expanding initial queries. Proposed algorithm is tested on OHSUMED test collection and Results show improvements in precision of the system compared to both keyword based and semantic approaches.

Future work in this area would be studying methods for extracting a number of additive concepts from different relations in MeSH ontology such as online notes, see also, annotation and pharmacological action which are neglected in our approach. In addition further studies for assigning appropriate weighting factors to concept pairs based on their frequencies over the corpus seems beneficial in this context. Yet another direction for further work would be selecting initial result set based on semantic approach or a combination of semantic and keyword based approaches, instead of using a predefined set or initial keyword based search results that are used in this experiment. The last direction for future work would be using global analysis for detecting co-occurrence of concepts in the whole corpus and comparing it with local analysis approach introduced in this paper.

References

[1] Aronson, A., Rindflesch, T.: Query Expansion Using the UMLS Metathesaurus. In: Proceedings of AMIA Annual Fall Symposium, pp. 485–489 (1997)
[2] Billerbeck, B., Zobel, J.: Techniques for Efficient Query Expansion. In: Apostolico, A., Melucci, M. (eds.) SPIRE 2004. LNCS, vol. 3246, pp. 30–42. Springer, Heidelberg (2004)
[3] Cui, H., Nie, J.J.: Query expansion by Mining user logs. IEEE Transaction on Knowledge and Data Engineering 15(4), 829–839 (2003)
[4] Gauch, S.: A corpus analysis approach for automatic query expansion and its extension to multiple databases. ACM Transactions on Information Systems (TOIS) 17(3), 250–269 (1999)
[5] Hersh, W., Price, S., Donohoe, L.: Assessing thesaurus-based query expansion using the UMLS metathesaurus. Journal of Americian Medical Informatics Association Annual Symp. (2000)
[6] Houston, A.L., Chen, H.C.: Exploring the use of concept spaces to improve medical information retrieval. Decision Support Systems 30(2), 171–186 (2000)
[7] Jalali, V., Borujerdi, M.R.M.: The Effect of Using Domain Specific Ontologies in Query Expansion in Medical Field. Innovations in Information Technology, 277–281 (2008)

[8] Leroy, G., Chen, H.C.: Meeting Medical Terminology Needs: The Ontology-enhanced Medical Concept Mapper. IEEE Transactions on Information Technology in Biomedicine 5(4), 261–270 (2001)
[9] Li, Y., Bandar, Z., McLean, D.: An Approach for Measuring Semantic Similarity between Words Using Multiple Information Sources. IEEE 45 Transactions on Knowledge and Data Engineering 15(4), 871–882 (2003)
[10] Mandala, R., Tokunaga, T., Tanaka, H.: Combining multiple evidence from different types of thesaurus for query expansion. In: SIGIR 1999: Proceedings of the 22nd Annual International ACM SIGIR, pp. 191–197 (1999)
[11] Mao, W., Chu, W.: Free-text medical document retrieval via phrase-based vector space model. In: Proceedings of AMIA Annual Symp., vol. 61(1), pp. 76–92 (2002)
[12] Rada, R., Bicknell, E.: Ranking documents with a thesaurus. Journal of the American Society for Information Science 40, 304–310 (1989)
[13] Srinivasan, P.: Query Expansion and MEDLINE. Information Processing and Management 32(4), 431–443 (1996)
[14] Voorhees, E.M.: Query expansion using lexical-semantic relations. SIGIR 94, 61–69 (1994)
[15] Yoo, S., Choi, J.: Improving MEDLINE Document Retrieval Using Automatic Query Expansion. In: Goh, D.H.-L., Cao, T.H., Sølvberg, I.T., Rasmussen, E. (eds.) ICADL 2007. LNCS, vol. 4822, pp. 241–249. Springer, Heidelberg (2007)

Using Connectors to Address Transparent Distribution in Enterprise Systems – Pitfalls and Options

Tomáš Bureš, Josef Hala, and Petr Hnětynka

Abstract. Software connectors are commonly used in component-based software engineering to model and implement inter-component communication. When used in a distributed environment, the responsibility of connectors is to make the remote communication transparent to components. For this task, connectors often employ a kind of middleware. In this paper, we evaluate the feasibility of making the remote communication completely transparent. We consider middleware commonly used in today's enterprise systems, namely CORBA and RMI in Java. We point out issues that prevent the transparency and analyze the impact on components together with possible tradeoffs.

1 Introduction

The high complexity of enterprise systems often leads to choosing distributed components as the key concept of the system architecture. In addition to reducing the complexity by providing encapsulated functional units which interact only via well interfaces, components also bring the power of reusability, which lies in distinguishing development of components from development of a system with components. In this separation, components are designed as independent functional units without concrete knowledge of their target environment. Only later they are assembled to form a particular system and they become deployed to a particular target environment.

The relative independence of a component on the target platform however requires special primitives (called *connectors*) to address the distribution — i.e. the

Tomáš Bureš, Josef Hala, and Petr Hnětynka
Department of Software Engineering, Faculty of Mathematics and Physics,
Charles University Malostranske namesti 25, Prague 1, 118 00, Czech Republic
e-mail: {bures, hala, hnetynka}@dsrg.mff.cuni.cz

Tomáš Bureš
Institute of Computer Science, Academy of Sciences of the Czech Republic
Pod Vodarenskou vezi 2, Prague 8, 182 07, Czech Republic

R. Lee, N. Ishii (Eds.): Software Engineering, Artificial Intelligence, SCI 209, pp. 81–92.
springerlink.com   © Springer-Verlag Berlin Heidelberg 2009

case when components are located in different address spaces or on physically different nodes. The typical goal of connectors is to make the distribution transparent to components by providing well-defined interface, while internally they rely on the use of middleware. In ideal case, this transparency allows making the decision about the exact distribution of components as late as during deployment time (i.e. just before components are instantiated on target nodes). Another important benefit of the idealistic use of connectors is the independence of particular communication middleware, which again allows its selection as late as in deployment time based for example on the operating system and libraries installed on the target nodes.

Our long-term experience with building component systems which employ connectors however shows that the reality is far from this idealistic view on connectors and that the completely transparent distribution, which connectors promise, is basically unattainable.

In this paper we show the main obstacles and pitfalls which prevent the connectors to completely hide the distribution. We also discuss how the pitfalls impact the component system and what restrictions and requirements they bring. For selected problems, we show how we have addressed them in our *connector generator* (i.e. a framework which is able to automatically create connector implementation based on a declarative specification). We base our discussion on two most prevalent middleware platforms, namely Java RMI [4] (Java Remote Method Invocation), CORBA [6] (Common Object Request Broker Architecture), and on Java platform, which is very typical in enterprise systems.

Structure of the text is as follows, Section 2 gives an overview of the software connectors and RMI and CORBA middleware. Based on this knowledge, Section 3 identifies main issues of transparent distribution. Section 4 relates these issues to component systems and shows their impact and our solution to them. Section 5 presents the related work, while Section 6 concludes the paper and outlines the future work.

2   Overview

2.1   Software Connectors

Software connectors are entities used to connect distinct components (more specifically their interfaces). During design they are represented as annotated inter-component bindings, which model the component interaction. Their annotations express requirements on the communication link (e.g. that a link must be secure, local). During runtime, connectors have form of code that takes care of the physical communication. The creation of a runtime connector based on its design counterpart is realized automatically during deployment with the help of *connector generator* [1].

From the modeling point of view, a connector is an inherently distributed entity (see Fig. 1). It consists of a number of *connectors units* (typically one server unit and a number of client units). The unit is linked directly to a component interface and the communication between them is realized by a local procedure call. The

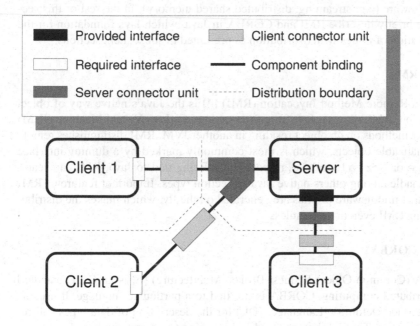

Fig. 1 A component application using connectors

communication among units is usually more complex as it often spans over several address spaces and relies on some sort of communication middleware.

The API between the component implementation and the connector unit is defined by the component system and it is parameterized by the business interface and design time annotations. This reflects the key idea of using connectors, which is that the component implementation depends on the component system, but it is unaware of how the communication between components is realized.

2.2   Middleware

By making components independent of how the communication is realized, the knowledge of the low-level details of the communication is pushed into connectors. When the communication spans several address spaces (as is the typically case), they internally use a kind of middleware. Each connector unit thus holds a middleware proxy. The role of the connector unit is to setup and instantiate the proxy and provide extra functionality not typically handled by the middleware (e.g. monitoring, adaptation). The connector unit in fact abstracts the middleware and forms a bridge between the component implementation and the particular middleware.

In the discussion in this paper, we focus mainly on RMI and CORBA middleware in the context of Java implementation, which reflects the needs of enterprise systems. In our general work on connectors, we however support a wide variety of middleware (e.g. streaming, distributed shared memory). In the rest of this section, we briefly describe RMI and CORBA in Java, which lays foundation for the identification of issues with distribution as presented in subsequent sections.

2.2.1   RMI

The Java Remote Method Invocation (RMI) [4] is the Java's native way of object distribution. RMI allows a Java program running in a Java Virtual Machine (JVM) to invoke methods on an object residing in another JVM. RMI distinguishes remote and serializable objects, which is most commonly marked by a dummy interface (`Remote` or `Serializable`, respectively). Being tied to Java, RMI can seamlessly handle among others native Java collection types. Important feature of RMI is also that middleware proxies are generated on the fly, which makes the distribution using RMI even more seamless.

2.2.2   CORBA

CORBA (Common Object Request Broker Architecture) [6] is an OMG standard for distributed computing. CORBA is not tied to a particular language. It uses its own Interface Definition Language (IDL) for the description of data types and remote interfaces. The IDL has standardized mapping to a large number of programming languages (including C, C++, Python, Ada, Java, List, etc.). The typical use of CORBA consists of specifying the remote interfaces and data types in IDL and then using an IDL compiler to build source code for proxies and serializable objects.

The platform independence of CORBA allows connecting clients and servers written in different programming languages. However, the tax for it is using CORBA-specific types, which are not well aligned with the native Java ones. This issue is mitigated by using RMI-IIOP technology [5], which allows generation of CORBA proxies based on the specification in Java (as opposed to starting with an IDL specification). This keeps the benefit of using mature and feature rich CORBA middleware, however the downside is the worse interoperability with other languages.

3   Issues of Transparent Distribution

As explained in previous sections, ideally the connectors have to provide a transparent distribution for component applications. But, in reality, the total transparency is not achievable. In this section, we enumerate a list of the most common issues that hinder from the existence of totally transparent distribution.

The issues are divided into three categories and use RMI and CORBA in Java to illustrate the problems.

3.1   *Limitations of Remote Interfaces and Transferable Objects*

Remote interfaces and transferable objects (i.e., objects used in remote communication) have usually additional requirements compared to those used only locally.

These limitations generally stem from an underlying layer (middleware) which implements and provides functionality in order the whole distributed application could actually work. Issues showed in this subsection are rather syntactical — objects and interfaces that play a role in remote communication need to follow a particular structure.

1. **Serialization of objects**

   The common requirement on transferable object is that they have to be serializable. Serialization is a process how objects are transformed into bit streams in order they can be restored later. A particular form of serialized data is not important (it can be in a binary or text form).

   In Java RMI, transferable objects have to implement the `java.io.Serializable` interface (otherwise, the RMI layer is not able to transfer the object and throws an exception). On the other hand, in CORBA, the transferable objects have to be either defined as structures (in case they represent only data without any methods) or as value types. Nevertheless, in both middlewares, such objects require special definition, which limits the transparent usage.

2. **Objects bound to a specific context/location**

   Objects can be closely tied with a context or computer where a particular component runs. These objects cannot be simply passed to another computer, even in case they are serializable.

   An example can be an object representing a file (e.g. in Java `java.io.File InputStream`) or network socket (`java.net.Socket`).

   In general, any object connected with an OS resource cannot be passed across the network.

3. **Specific structure of remote interfaces**

   Each language/middleware has different support for remote computing and its own specific limitations, constructs and standards to declare and use remotely accessible interfaces.

   For example in Java RMI, a remotely accessible interface has to extend `java.rmi.Remote` interface and each its method has to declare `java.rmi.RemoteException` exception. In addition, when RMI-IIOP is used, there are limitations laid on the names of methods in order they can be transformed into CORBA IDL (e.g. overloading of methods should be avoided as IDL does not support it).

3.2 Different Behavior between Local and Remote Objects/Interfaces

Besides the syntactical limitations (described above), a developer has to be aware of possibly different behavior of class fields and local/remote calls.

4. **Static fields**

   Remote objects should avoid using of static fields because they are not shared between different address spaces.

In a single address space, static fields can be seen as global variables available from everywhere. But it does not work for applications consisting of more components running in different address spaces, because there will be a different copy of the static field in each address space.

5. **Timing**
   Local and remote calls have different calling times (remote calls take longer time). For specific application parts (e.g. user interface), it is essential how fast the call are proceeded and the overhead of the call itself has to be minimal.

6. **Passing arguments by value vs. by reference**
   In particular languages, argument passing can have different semantics between local and remote calls. In Java, objects are in method calls passed by reference. But in the case of remote call, the arguments are passed by value (and therefore the changes made on a server are not visible on a client).

7. **Pointers**
   The pass-by-value semantics of arguments means that it is necessary to perform deep copy for data structures, which internally use pointers. In addition to the problem of two unrelated copies existing on client and server (as discussed above). There is also a problem that the deep copy often results into transfer of large data structures. For example when one vertex of a graph is passed to the server the whole graph structure must be transferred from the client to the server. On top of it, there are middlewares (e.g. SOAP), which do not support pointers at all.

8. **Failure of remote calls**
   When an application does not know that the method calls can be remote, it does not assume that the calls can fail due to network problems. The exception is unexpected and it is a question how the application should react to the exception.

3.3   Issues of Connector Implementation

The previous issues are related to component applications and remote-procedure call (RPC) in general. This subsection further discusses various implementation problems associated specifically when using middleware in connectors.

9. **Dealing with remote references**
   Middlewares typically support passing of remote references which can be handed between application components (they can be passed as method arguments or returned from the calls). When connectors are used, they in fact substitute the remote references and therefore each remote reference passed in a method call has to be automatically wrapped in a connector.

10. **Conversions between native and data types of used middleware**
    Middlewares commonly limit the available data types, which can be used in remote calls. In order that the connectors can make communication transparent, application data types have to be transparently mapped onto middleware data types and vice-versa. But such a mapping greatly impacts the performance of the

calls. On the other hand, restrictions laid on the component interfaces (in terms what types may be used such as mapping to middleware is simple enough) impair the middleware transparency.

11. **Heterogeneous type systems**
    The clash between native and middleware data types becomes even more problematic when a component interface is to be made accessible by two different communication middlewares (e.g. RMI and SOAP).

4 Dealing with the Issues in Component Systems

Above, we have discussed issues hindering an existence of fully transparent connectors in component systems. In this section, we show impacts of these issues on the component systems together with their potential solutions (provided the particular issue can be overcome at all). The section is divided into several subsection based on the kinds of impacts. Since, the issues are grouped with respect to components, they are ordered differently compared to Section 3. However the related issues are always explicitly enumerated at the beginning of each subsection.

4.1 Different Types

Connectors try to shield applications from communication middleware and its types. In the ideal situation, applications should have no notion of the remote distribution and therefore there should not be any constraints on used types. However, this is not feasible, as shown by issues 1 – *Serialization of objects*, 3 – *Specific structure of remote interfaces*, 10 – *Conversion between native and data types of used middleware*, and 11 – *Heterogeneous type systems*.

The common denominator of these issues is that either the middleware type system has to be known and used by the the component implementation (which goes against the aim of using component and connectors) or the types and interfaces of the application have to be mapped to middleware specific ones. The other option is also not very viable, as it involves relatively big overhead (connected with conversion between the types) and it also requires information about how types can be serialized. Finding a balanced tradeoff thus seems to be the best possible solution. Moreover, when dealing with connector generation, it is necessary that the generator can handle several middlewares, which means it has to be indifferent to a particular type system, at least to an extent.

In our connector generator, we have addressed these issues by introduction of an internal universal type system (called connector type system). This type system allows describing types in general, but most importantly it is invisible to users of the generator.

The connector type system defines three main concepts — a primitive type, interface, and array of types. These concepts are extended for an actual type system (e.g. the current implementation allow to use Java and Corba types). In addition, the connector type system provides operators, which allow performing various operations

on the types. E.g. java_interface('IFace') will result into a Java interface
corresponding to the IFace type; rmi(java_interface('IFace')) will result into a Java interface corresponding to the IFace type but this time enhanced
by definitions necessary to use the interface as remote via RMI.

Via the connector type system and operators (which can be easily added into the
generator), conversions among types can be easily performed.

With the help of the connector type system, the connector generator can automatically serialize and de-serialize objects passed as parameters. However, such an
automatic serialization of all parameters is not sound idea, as an object could be tied
with context/computer (the issue 2). Another reason may be that an object could
contain private and sensible information. If such an object was serialized and sent
through the network without awareness of the user, it would be a security hole.

Therefore, our connector generator does not perform the automatic serialization
of all parameters implicitly, but requires their explicit specification. For applications
written in Java, we rely on the language native concept, which is the Java-based
serialization.

For dealing with remote interfaces, the connector type system offers a set of operations, which allow enhancing the interfaces with necessary definitions in order to
make their remote usage possible. An example is the above mentioned rmi operator, which for a Java interface prepares its remote version, i.e. it modifies the original
interface to subclass from java.rmi.Remote and for each its method it declares
java.rmi.RemoteException. For example, the interface

```
public interface Hello {
   String message(String text);
}
```

is transformed into

```
public interface Hello
          extends java.rmi.Remote {
   String message(String text)
          throws java.rmi.RemoteException;
}
```

The conversion between type systems is however a complex task, which is not
always possible. When supporting both CORBA and RMI a common issue is converting between IDL and Java.

In the direction from IDL to Java, it is possible to use any IDL compiler targeting
Java, e.g. idlj (bundled with recent Java Standard Edition). However, the IDL
interface can be quite complex and can use many types defined in the IDL file.
Moreover, the IDL file itself may include other IDL files. Thus a generation of a
corresponding Java interface can result in creation of many types which the interface
depends on and which are difficult to relate to the types used in the application. The
opposite direction (from Java to IDL) may be to some extent realized in a similar
way using the RMI compiler (i.e. rmi). However, the same limitations and problems
apply.

The conversion between type systems thus works reliably only for simple interfaces with primitive data types, which can be directly mapped to Java types. For any other interfaces, the component design has to reflect the use of a particular middleware.

4.2 Different Behavior of Local and Remote Calls

As shown in issues 5 – *Timing*, 6 – *Passing arguments by value vs. by reference*, 7 – *Pointers*, 8 – *Failure of remote calls*, the type-system is not the only thing that changes with distribution; there is also a difference in the semantics of local and remote calls.

The time a remote call takes is quite unpredictable and depends on a number of aspects. Although, the unpredictability may be to a certain point mitigated by the use of middleware with real-time guarantees, there is no way how to decrease the latency of distributed calls under a particular threshold. This means that a component application cannot be distributed arbitrarily, but only on interfaces where higher latency does not jeopardize the correctness and usability of the application. To mark such interfaces, interface annotations may be used (with values such as *local*, *LAN*, *WAN*). These annotations are used later by the deployment planning tool and connector generator to ensure that the application is not wrongly distributed.

While the issue of timing relates to the different latency of distributed communication, the issues of argument passing and pointers are related to throughput. The problem is that wrongly distributed application may tend to transfer large datastructures. This gets visible when the network connection between target nodes is not fast enough. As opposed to latency, the throughput may be typically increased by new hardware. The options of addressing the issues in component systems are similar to those of the timing — marking interfaces where the distribution boundary is permissible and where it is not.

The distribution also brings the issue that the remote call may fail due to network problems. However, this is not a serious issue, as when connectors are dealt explicitly, it is reasonable to expect that a failure in a connector may happen — regardless whether local or remote communication is used. On the other hand, connectors may hide some transient errors via timeouts and retries, which may be parameterized in a connector specification.

4.3 Sharing Variables

The common feature of issues 2 – *Objects bound to a specific context/location*, 4 – *Static fields*, and partly of 6 – *Passing arguments by value vs. by reference* and 7 – *Pointers*, is that data is shared between two entities. This is however forbidden in component systems as it violates the properties of encapsulation and communication only via well-defined interfaces. Thus, these issues do not constitute a problem by themselves, rather their manifestation in a component application points to more serious design flaws.

4.4 Remote References

The issue 9 – *Dealing with remote references*, requires that a connector unit (or a whole connector) is created when a reference to a component interface is passed as a parameter. This is necessary for ensuring that every component interaction happens through component interfaces and a connector that binds them. Achieving this behavior is generally possible, however a few technical obstacles have to be overcome. These among others comprise creating transmitting remote references, identification of them. In this subsection, we describe how we have addressed the issue in our connector generator.

To give a background, Figure 2 illustrates the differences between the local (the `non_remote` variable) and remote (the `remote` variable) references when using communication middleware. The local reference is in fact a pointer directly to the target object, while the remote reference is a pointer to a special object, called *client proxy*. The client proxy is local and implements the same interface as the remote object (in this case the `DemoIFace` interface) but the methods do not implement the business logic but the logic of passing arguments over the network. On the target side, there is an object called *server proxy* that implements receiving the calls from the stub and calls the real object. After the call is finished, the returning value is again passed via the skeleton and stub.

When connectors are used, the client and server proxies are part of the connector. Thus, the component has a local reference to the connector, which internally encapsulates the client proxy. When a component passes a reference to another component, it thus passes the reference to the client connector unit.

When such a reference is transmitted, the connector unit on the originating side translates the reference to a special object called wrapper. The wrapper contains all information to establish the link between the referenced interface and the side which will receive the reference. Further it is serializable, so as it may be passed by value and it implements all the interfaces of the reference, such as it can be used inside complex types. The implementation of the methods is empty, as it is an error to call a method on the wrapper.

The connector unit on the receiver's side, then creates a new connector unit based on the wrapper and forms a connector that captures the new binding.

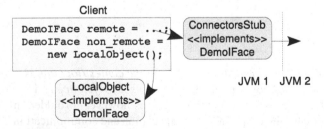

Fig. 2 The difference between a remote reference and a local reference

To specify, what wrapper class is used for a particular type of reference, we use Java annotations. Further, we use Java annotations to distinguish wrapper classes from component interfaces (i.e. the "real" references).

A special attention in identification of remote references has to be paid to a possibility of circular references — an object can hold (directly or indirectly) a reference to itself. Our implementation also supports the circular references.

5 Related Work

The idea of some unification of middlewares and creation of an abstract layer is presented also in the project Arcademis [7]. It aims at object oriented middleware and the main purpose why the Arcademic project was founded is that current convectional object oriented middlewares are monolithic and inflexible to meet the needs of modern rapidly changing technologies. Nowadays there exist many different devices like cell phones, PDAs, etc., with various demands on quality and characteristics of a connection. Arcacemis tries to address the limitations of current middleware implementations. It is designed to be flexible, so that new transport protocols, connection management policies and authentication algorithms can be easily configured. But Arcademis in only a "template" how some middleware could look like, it does not implement any factual functionality.

The other project using various middlewares for providing inter-component communication is PadicoTM [3]. It aims at parallel computational infrastructures called the grid. The grid encourages development of new applications in the field of scientific computing. It is used for example for simulation of complex physical processes. Different components of a scientific application may demand different communication paradigms like RMI (Remote Method Invocation) or MPI (Message Passing Interface). The aim of PadicoTM is to allow several communication middleware and runtime to efficiently share network resources and use the most suited communication for a specific application.

6 Conclusion and Future Work

In this paper, we have identified issues that come in play when striving for transparent distribution using connectors. The common characteristics of the problems is that the use of middleware enforces syntactical restrictions on the interfaces and data types and changes non-functional properties of the communication (e.g. different execution times as opposed to the local call). We have evaluated these problems from the point of view of component-based software engineering and discussed possible options and tradeoffs. The solutions presented in the paper have been implemented and verified in the frame of SOFA 2 component system [2, 8].

The middleware discussed in the paper has targeted RPC in enterprise systems (i.e. RMI and CORBA in Java). In the future work, we would like to focus on additional communication styles (e.g. messaging, streaming, distributed shared memory) and different application domains (e.g. embedded systems).

Acknowledgements. This work was partially supported by the Czech Academy of Sciences project 1ET400300504.

References

1. Bures, T., Malohlava, M., Hnetynka, P.: Using DSL for Automatic Generation of Software Connectors. In: Proceedings of ICCBSS, Madrid, Spain, pp. 138–147. IEEE Computer Society Press, Los Alamitos (2008)
2. Bures, T., Hnetynka, P., Plasil, F.: Runtime Concepts of Hierarchical Software Components. International Journal of Computer & Information Science 8(S) (2007)
3. Denis, A.: Meta-communications in component-based communication frameworks for grids. Journal of Cluster Computing 10(3) (2007)
4. Java Remote Method Invocation, http://java.sun.com/javase/6/docs/technotes/guides/rmi/index.html
5. Java RMI over IIOP, http://java.sun.com/javase/6/docs/technotes/guides/rmi-iiop/index.html
6. OMG. Common Object Request Broker Architecture Specification, OMG document formal/2008-01-04 (2008)
7. Pereira, F.M.Q., Valente, M.T.d.O., Pires, W.S., da Silva Bigonha, R., da Silva Bigonha, M.A.: Tactics for Remote Method Invocation. Journal of Universal Computer Science 10(7) (2004)
8. SOFA 2, http://sofa.objectweb.org/
9. Sun Microsystems, JSR 220 – Enterprise JavaBeans 3.0 (2006)
10. Szyperski, C., Murer, S., Gruntz, D.: Component Software, 2nd edn. Beyond Object-Oriented Programming. Addison-Wesley, Reading (2002)

An Empirical Research on the Determinants of User M-Commerce Acceptance

Dong-sheng Liu and Wei Chen

Abstract. This paper proposes an integrated model to investigate the determinants of user mobile commerce acceptance on the basis of innovation diffusion theory, technology acceptance model and theory of planned behavior considering the important role of personal innovativeness in the initial stage of technology adoption. The proposed model was empirically tested using data collected from a survey of MC consumers. The structural equation modeling technique was used to evaluate the causal model and confirmatory factor analysis was performed to examine the reliability and validity of the measurement model. Our findings indicated that all variables except Perceived risk and perceived ease of use significantly affected users' behavioral intent.

1 Introduction

With rapid development of mobile technology and emergence of 3G, more and more enterprises invest heavily in the new technology application of mobile commerce. According to the report of Gartner Group (2001a), more than 80 percent of European enterprises agree that mobile devices and mobile commerce application are momentous to business management. About 70 percent of 3500 enterprises in the world are taking mobile data exchange into consideration and 20 percent of them have already managed to employ this application [1]. In 2005, the investigation of Yankee Group shows that in US mobile commerce users reaches about 50 million, near 40 percent of all the workers. These mobile users eagerly expect get work outside the offices completed efficiently with mobile technology [2].But the 2006 report on China's mobile commerce from Eriel Advisory shows that in 2005, China has more than 3 million mobile commerce users, just taking up 3 percent of all the mobile users in the world and less than 1 percent of mobile users in the corresponding period. The user mobile commerce acceptance is significantly insufficient [3]. Therefore, deep investigations on the determinants and mechanism of user mobile commerce acceptance will facilitate the application of mobile commerce.

Dong-sheng Liu and Wei Chen
College of Computer Science & Information Engineering
Zhejiang Gongshang University, 310018, Hangzhou, China
e-mail: lds1118@mail.zjgsu.edu.cn, true1.chen@gmail.com

R. Lee, N. Ishii (Eds.): Software Engineering, Artificial Intelligence, SCI 209, pp. 93–104.
springerlink.com   © Springer-Verlag Berlin Heidelberg 2009

2 Basic Concepts, Research Model and Hypotheses

In the previous researches on information technology acceptance, some theories can to some extent explain and predict user technology acceptance behavior, but no theory cover all the determinants of technology acceptance (e.g. TAM contains users' perception, TPB focuses on external environment including organization, IDT emphasizes the innovation characteristics of technologies.), which makes models have unsatisfactory explanations. These models can't comprehensively predict and explain user technology acceptance behavior. Legris (2003) argued that researches on information technology acceptance behavior should not be limited. Integrating different models for comprehensive analysis is necessary. The author does not propose specific model further [4]. Venkatesh (2003) integrates TAM and TPB and proposes a unified theory of acceptance and use of technology model (UTUAT). However, this unified model can't explain the acceptance behavior of this specific technology application of mobile commerce [5]. In addition, previous researches on technology acceptance focus on users' perception, e.g. perceived ease of use and perceived usefulness. But in the initial stage of technology adoption, adopters have little knowledge on the characteristics of technologies and external environments. Under this circumstance, biographical characteristics (e.g. personal innovativeness) are more important to technology acceptance behavior [6], but few researches involve the impact of personal innovativeness on user technology acceptance. This paper proposes an integrated model of user mobile commerce acceptance by integrating perceived ease of use and perceived usefulness from TAM, subjective norm and perceived behavior control from TPB, innovativeness from IDT and perceived risk.

2.1 Technology Acceptance Model (TAM)

Davis (1989) revises TRA and proposes technology acceptance model, TAM [7]. TAM argues that a certain MIS adoption depends on perceived usefulness (the extent to which MIS improves work performance) and perceived ease of use (the extent to which efforts are needed for MIS). External variables (e.g. training) influence behavioral intention to use, with perceived usefulness and perceived ease of use as mediators. TAM is widely used in information technology acceptance for its simple structure and satisfactory explanations. Some scholars have different arguments. Mathieson (1991) points out that TAM covers general determinants of technology acceptance and needs support of other models and theories to better prediction of user technology acceptance behavior [8].

In previous researches on technology acceptance model, the perceived usefulness, perceived ease of use and behavioral intention to use are stable and consistent. Behavioral intention to use is mainly influenced by perceived usefulness and perceived ease of use while perceived ease of use influences perceived usefulness. Thereby, this paper hypothesizes:

- H1a: perceived ease of use is positively related to behavioral intention to use mobile commerce.
- H1b: perceived ease of use is positively related to perceived usefulness.
- H2: perceived usefulness is positively related to behavioral intention to use mobile commerce.

2.2 Theory of Planned Behavior (TPB)

Ajzen argues that positive attitude and subjective norm will not bring intentions or behavior because of insufficient resources and abilities. So Ajzen (1991) proposes theory of planned behavior (TPB) in order to appropriately predict and explain personal behaviors on the basis of theory of Reasoned Action (TRA) [9]. Comparing with TRA, TPB has an additional variable of perceived behavior control.

Perceived behavior control refers to individual perception of necessary resources for certain behavior. TPB argues that perceived behavior control determines intention and behavior. Joo (2000) founds that whether suppliers provide good training for employees and periodical device maintenance is significantly correlated with IT adoption of CEOs through conducting an empirical research on the determinants of IT adoption of CIOs in 162 SMEs [10]. Because mobile commerce is a new technology application, most of users have no previous experience. Necessary guidance and training provided to users before can enhance users' self-efficacy and is helpful to promote users' intention to use. Therefore, this paper hypothesizes:

- H3: Perceived behavior control is positively related to behavior intention to use mobile commerce.

Subjective norm refers to the extent which people around support users in adopting new technologies, similar to social factors or social influences. Many scholars study impact of subjective norm on behavior intent and have different conclusions. Davis (1989) finds that the correlation between subjective norm and behavior intent is not significant. Therefore, he removed subjective norm form TAM [7]. But Taylor and Todd (1995) find the correlation between them is significant [11]. The reason why different conclusions come out because they overlook that new technology adoption is voluntary or compulsory. User mobile commerce adoption this paper studies is independent behavior, with no coercion. At the same time, mobile commerce is beginning to develop. Most of users have no previous experience. So, subjective norm greatly influence users' information system perception. This paper hypothesizes:

- H4: Subjective norm is positively related to behavior intention to use mobile commerce.

2.3 Innovation Diffusion Theory (IDT)

Innovation Diffusion theory originated from sociology demonstrates that the main determinant of innovation diffusion is perceived innovation characteristics (Rogers, 1993) [12]. Rogers in his research proposes five innovation characteristics: relative advantage, compatibility, complexity, trial ability and observables and further proposes individual factor of innovation diffusion-personal innovativeness. Innovation adoptions are classified according to it. Agarwal and Prasad (1998) initially adopt personal innovativeness as one indicator in information technology, which is taken as an antecedent of perceived information technology [13].

Innovativeness refers to the tendency which individuals or other innovation units accept new technologies earlier than other members. Rogers (1983) proposed five adopter categories: innovators, early adopters, early majority, late majority and laggards according to the tendency of new technology adoption [12].

Previous researches on technology acceptance demonstrate the personal innovativeness influences subjective perception and reflects the tendency of which individuals intend to adopt information technology. Innovativeness has a positive effect on the correlation between technology perception and technology acceptance. In addition, in the innovation diffusion process, innovation cognition involves subjective factors and social factors. Perceptions are influenced by interpersonal networks of adopters in addition to the influence of innovation spreaders. Technology acceptance has its given social system. Structure and norm of social system becomes important background for innovation adoption. Conformity of new technologies and social value will promote new technology acceptance. Therefore, this paper hypothesizes:

- H5: Personal innovativeness is positively related to subjective norm.

IDT argues that individuals with high personal innovativeness tend to accept new opinions and changes and are more capable of coping with uncertainty and risks. Roger argues that early adopters usually have higher economic and social status than late adopters. To some extent this ensures that early adopters have sufficient resources to use new technologies and are willing to take certain risk.

Rogers (1983) finds that early adopters have advantages over late adopters in abstract information processing, rationality, decision-making, risk and uncertainty avoidance, usually with more knowledge and intelligence through conducting a comparative research on personality and abilities of early adopters and late adopters. Early adopters seldom consider the difficulties in using new technologies. Early adopters are risk-taking and in pursuit of excitement, which makes them overlook the difficulties [12]. Users with high innovativeness tend to think that benefits from time and cost savings outweigh risks in using mobile commerce. Most of them have much IT experience, breakdown handling skills and Trojan horse and virus preventing techniques.

Early adopters clearly know that their adoption is important to obtain respect from others and keep them in the center of communication networks of social system. High status can help them work more efficiently. Yi (2006) finds that one of most important reasons is that using PDA conveying status can create a favorable environment through conducting an empirical research of 222 surgeons [14]. Before mobile commerce is widely used, some users think that new mobile terminal devices convey fashion and fortune. Adopting these devices promotes personal images and indirectly improves work performance. Thereby, this paper hypothesizes:

- H5b: Personal innovativeness is negatively related to perceived risk.
- H5c: Personal innovativeness in positively related to perceived ease of use.
- H5d: Personal innovativeness in positively related to perceived usefulness.
- H5e: Personal innovativeness in positively related to perceived behavior control.

2.4 Perceived Risk

Risk refers to uncertainty of networks and trading systems, that is, possible failure, divulgence, steal in trade process or economic risks because of unexpected results. Data from CNNIC shows that lack of safety and privacy protection makes 85 percent of netizens never use e-commerce [15]. Mobile commerce as one type of e-commerce has similar risks. So, as long as users think that mobile commerce is low risk, they will accept mobile commerce. Therefore, this paper hypothesizes:

- H6: Perceived risk is negatively related to behavior intention to use mobile commerce.

2.5 Research Model

According to the preceding hypothesis, the determinants of user mobile commerce acceptance model as Fig. 1 presents:

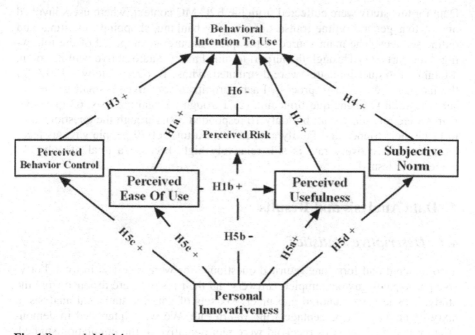

Fig. 1 Research Model

3 Research Methodology

3.1 Measurement Development

Previous research was reviewed to ensure that a comprehensive list of measures were included. Those for perceived usefulness, perceived ease of use, perceived risk were adapted in our model from previous studies on a revised TAM [16]. The construct for behavioral intention to use, Subjective Norm were adapted from the studies of Venkatesh et al. [5]. The scales for Personal innovativeness were based on J.Lu, J.E Yao and C.S.Yu [6]. The measures for behavioral intention to use were captured using three items derived from Taylor, S. & Todd, P. A[11].

The survey questionnaire consisted of two parts. The first recorded the subject's demographic information. The second recorded the subject's perception of each variable in the model. The demographic variables assessed were gender, age, level of education, frequency of using a cellular phone and the degree of familiarity with using online services. The second section asked each subject to indicate his or her degree of agreement with each item. Data were collected using a five point Likert-type scale, where 1 indicated strongly disagree; 2 showed disagreement to some extent; 3 stood for uncertain; 4 was for agree to some extent; and 5 indicated strong agreement.

3.2 Subjects

Data for this study were collected from the B2C MC context, where users invoked one of four general online transactions: online banking, shopping, investing, and online services. The main source of the questionnaire is composed of the following three parts: (1) Through the airport terminal and the automotive waiting room, a total of 146 questionnaires were distributed, whose recovery rate was 100% for the face-to-face exchange process. Each participant was given a small gift to ensure the validity of the questionnaire. (2) Through the mailing lists, 63 questionnaires were randomly sent, but only 16 responded. (3) Through the questionnaires to bring their friends and family around them, a total of 79 people gave us feedback, and its recovery rate is still relatively high, because a total of only 114 copies were issued.

4 Data Analysis and Results

4.1 Descriptive Statistics

Two hundred and forty-one returned questionnaires were received in total. Forty-five participants gave incomplete answers and their results were dropped from the study. This left one hundred and ninety-six sets of data for statistical analysis, a sixty point six eight percentage valid return rate. We were interested in demonstrating that the responses received were representative of the population studied. A multivariate analysis of variance was therefore undertaken to determine whether differences in response time (early versus late) were associated with different response profiles. The results indicated no significant difference in any of the variables of interest. Tables 1 list the sample demographics. The data indicates that the majority of the respondents had a college education. Ninty-one point eight four percent of them were between the ages of 20–39. Respondents aged 20- 29 account for 71.94 percent, which shows that most of netizens in China are young men. 196 respondents consist of 115 men taking up 58.67 percent and 81 women,

Table 1 Demographic attributes of the respondents

Item	Frequency	Percent	Cumulative Percent
Gender			
Male	115	58.67	58.67
Female	81	41.33	100.00
Education level			
senior high school and under	33	16.84	16.84
junior college	48	24.49	41.33
undergraduate	100	51.02	92.35
Graduate (or above)	15	7.65	100.00
Age			
Less than 20	10	5.10	5.10
20-29	141	71.94	77.04
30-39	29	14.80	91.84
40-49	29	6.12	97.96
Over 50	4	2.04	100.00
Use cellular phone			
No cellular phone	4	2.04	2.04
Use in urgent need	10	5.10	7.14
Use casually	25	12.76	19.90
Use for convenience only	97	49.49	69.39
Use frequently	60	30.61	100
Familiar with online transactions			
Completely unfamiliar	39	19.89	19.89
Familiar a little	87	44.39	64.28
Neutral	31	15.82	80.10
Familiar	27	13.78	93.88
Very familiar	12	6.12	100

41.33 percent. The sex ratio is in accordance with the data from Statistical Report on the Internet Development in China released by CNNIC, which shows this data is representative [15].

4.2 Measurement Model

The proposed research model was evaluated using structural equation modeling (SEM). The data obtained were tested for reliability and validity using confirmatory factor analysis (CFA). This step was used to test if the empirical data conformed to the presumed model. The model included 27 items describing seven latent constructs: perceived ease of use, perceived usefulness, perceived behavior control, perceived risk, Subjective Norm, Personal innovativeness, and behavioral intention to use. The CFA was computed using the LISREL software. The measurement model test presented a good fit between the data and the proposed measurement model. For instance, the chi-square/degrees of freedom (1830.684/465) were used because of the inherent difficulty with sample size. The x2/d.f. value was 3.25, among Joreskog and Sorbom's suggestion [17] between two and five [18]. The model has a good fit to the data, based on this suggestion. The various goodness-of-fit statistics are shown in Table 2. A GFI is 0.9. RMSEA was slightly greater than the recommended range of acceptability (<0.05–0.08) suggested by MacCallum et al. [17]. Thus, the measurement model has a good fit with the data, based on assessment criteria such as RMR, GFI, NFI, NNFI, CFI and RMSEA.

The composite reliability was estimated to evaluate the internal consistency of the measurement model. The composite reliabilities of the measures included in the model ranged from 0.83 to 0.93 (see Table 3). All were greater than the benchmark of 0.60 recommended by Bagozzi and Yi [19]. This showed that all measures had strong and adequate reliability and discriminate validity. As shown in Table 3, the average variance extracted for all measures also exceeded 0.5. The completely standardized factor loadings and individual item reliability for the observed variables are presented in Table 4.

Fig. 2 presents the significant structural relationship. Among the research variables and the standardized path coefficients, Most of the hypotheses were strongly supported except for hypotheses H1a, H5b and H6. The perceived ease of use effect on behavioral intention to use MC is not significant (H1a). While, it does have

Table 2 Model evaluation overall fit measurement

Measure	Value
Root mean square residual (RMR) (<0.05)	0.03
Goodness of fit index (GFI) (>0.9)	0.91
Normed fit index (NFI) (>0.9)	0.95
Non-normed fit index (>0.9)	0.93
Comparative fit index (>0.9)	0.94
Root mean square error of approximation (RMSEA) (<0.05-0.08)	0.04

Table 3 Assessment of the construct reliability

Variables	The composite reliability (>0.6)	Average variance extracted (>0.5)
perceived ease of use	0.86	0.74
perceived usefulness	0.93	0.68
perceived behavior control	0.88	0.64
perceived risk	0.91	0.66
Subjective Norm	0.92	0.67
Personal innovativeness	0.83	0.61
behavioral intention to use	0.88	0.69

Table 4 Standardized factor loadings and individual item reliability

Item	Measure/ References		Factor loading	R² >0.5
PEOU1	I think learning to use MC is easy.		0.815	0.723
PEOU2	I think finding what I want via MC is easy.	16	0.823	0.742
PEOU3	I think becoming skillful at using MC is easy.		0.856	0.817
PEOU4	I think becoming skillful at using MC is easy.		0.843	0.775
PU1	Using MC would improve my performance in online transactions.		0.663	0.542
PU2	Using MC would increase my productivity in online transactions.		0.878	0.786
PU3	Using MC would enhance my effectiveness in online transactions.	16	0.878	0.786
PU4	Using MC would make it easier for me to engage in online transactions.		0.716	0.727
PU5	I think using MC is very useful for me to engage in online transactions.		0.670	0.595
PR1	I think using MC in monetary transactions has potential risk.		0.818	0.843
PR2	I think using MC in product purchases has potential risk.	16	0.787	0.811
PR3	I think using MC in merchandise services has potential risk.		0.766	0.697
PR4	I think using MC has some other potential risk.		0.756	0.641
PBC1	I have the Mobile Terminal necessary to use the MC.		0.792	0.793
PBC2	I have the knowledge necessary to use MC.		0.707	0.764
PBC3	The Mobile Terminal compatible with other Business software.	5	0.832	0.846
PBC4	I can settle the problems easily when I encounter with difficulties in using MC.		0.684	0.753
SN1	People who influence my behavior think that I should use MC.		0.860	0.739
SN2	People who are important to me think that I should use the system.	5	0.849	0.718
SN3	In general, the organization has supported the use of MC.		0.789	0.674
PI1	If I heard about a new information technology, I would look for ways to experiment with it.		0720	0.692
PI2	Among my peers, I am usually the first to explore new information technologies.	6	0.729	0.707
PI3	I like to experiment with new information technologies.		0.697	0.584
PI4	In general, I am hesitant to try out new information technologies.		0.716	0.683
BI1	I intend to use MC in the nest (n) months.		0.800	0.843
BI2	I predict I would use MC in the next (n) months.	11	0.660	0.781
BI3	I plan to use MC in the next (n) months.		0.631	0.752

a significant indirect effect on behavioral intention to use through perceived MC usefulness, and also has a significantly positive direct effect on perceived MC usefulness (H1b: g=0.339, P<0.001). The data also shows that perceived usefulness-significantly directly influences the behavioral intention to use (H2: g= 0.384, P<0.001). For hypothesis H3, Subjective Norm has a significantly directly influences effect on the behavioral intention to use MC (H3: g=0.309, P<0.001). The Subjective Norm has a significantly directly influences effect on the behavioral intention to use MC (H4: g=0.391, P<0.001). For hypothesis H5b, The Personal innovativeness effect on perceived risk is not significant. However, it does have a significant indirect effect on Subjective Norm (H5a: g=0.105, P<0.001), perceived ease of us (H5c: g=0.140, P <0.001), perceived usefulness (H5d: g=0.129, P<0.001), perceived behavior control (H1e: g=0.069, P<0.001). The result confirms that the perceived risk effect on the behavioral intention to use MC is not significant (H6).

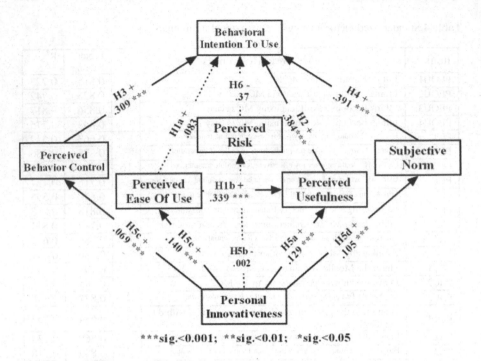

***sig.<0.001; **sig.<0.01; *sig.<0.05

Fig. 2 The empirical results of this study

4.3 Discussions

The results from the hypotheses testing are:

- Personal innovativeness is not closely related with perceived risk but is directly proportional to perceived ease of use, perceived usefulness, perceived behavior control, and subjective norm, among which perceived ease of use is the most influential. Personal innovativeness affects consumers' desire to use M-commerce indirectly by its influence on the above-mentioned variables. Therefore, in the primary stage, it is preferable to introduce M-commerce to people with strong personal innovativeness, and then they can make use of their social networking to diffuse to those who are with low personal innovativeness (the role of subjective norm). Besides, M-commerce suppliers should pay careful attention to the affect of system usefulness to consumers' acceptability. It is advisable to give wide publicity to system usefulness in order to enhance consumers' perceived usefulness.
- The research results indicate perceived usefulness is directly proportional to behavior intention to use, while perceived ease of use affects behavior intention indirectly by its influence on perceived usefulness. Therefore, M-commerce suppliers should try to simplify the system to ease the use of consumers to further stimulate their desire to use the system.

- In the empirically test, the perceived risk has not been supported as expected due to two reasons. Firstly, the majority of the subjects in this research have received college education, accounting for 83.16%. And the age of most subjects ranges from 20 to 29, accounting for 71.94%. Generally speaking, the young and well-educated are capable of hedging risks. Besides, it is also possible that the young usually have weak consciousness of avoiding risks. Secondly, the risks of M-commerce are objective while the perceived risk is subjective. Different subjects who hold different values have different opinions about the same thing. That's why subjective opinions can not reveal objective facts in some occasions.

5 Conclusions

In China, as Internet has been widely used and portable equipments have been developing rapidly, M-commerce has been attracting more and more consumers, who are getting more and more involved in. In this case, much attention has been paid to how to encourage consumers to accept and use M-commerce. This paper has carefully studied the process of how consumers accept M-commerce, with the aim of building a model, which can reveal the affects of different factors to consumers' acceptability. The paper has also justified the model by the combination of qualitative and quantitative analysis, the combination of positive and normative analysis, together with statistical methods. Therefore, the model has been proved to be able to explain, predict and manage the consumers' behavior of accepting.

Acknowledgments. Dong-sheng Liu thanks the China National Science Foundation (70671094), Zhejiang Technology Project (2008C14061), Foundation of National Doctor Fund(20050353003), Wei Chen thanks the Zhejiang xinmiao Plan (2008R40G2050029).

References

[1] Alanen, J., Autio, E.: Mobile business services: A strategic perspective. Idea Group Inc., USA (2003)
[2] Yankee Group: Mobile workers number almost 50 million. Business Communications Review 35(8), 8–12 (2005)
[3] China Wireless Application Research Report 2006, iResearch Consulting Group (2006)
[4] Legris, P., Ingham, J., Collerette, P.: Why do people use information technology? A critical review of the technology acceptance model, Information and Management 40(3), 191–205 (2003)
[5] Venkatesh, V., Morris, M.G., Davis, G.B., Davis, F.D.: User acceptance of information technology: toward a unified view. MIS Quarterly 27(3), 425–478 (2003)
[6] Lu, J., Yao, J.E., Yu, C.S.: Personal innovativeness, social influences and adoption of wireless Internet services via mobile technology. Journal of Strategic Information Systems (14), 245–268 (2005)

[7] Davis, F.D.: Perceived usefulness, perceived ease of use, and user acceptance of information technology. MIS Quarterly 13(3), 319–339 (1989)

[8] Mathieson, K.: Predicting user intentions: comparing the technology acceptance model with the theory of planned behavior. Information Systems Research 2(3), 173–191 (1991)

[9] Ajzen, I.: The theory of planned behavior. Organizational Behavior and Human Decision Processes 50(2), 179–211 (1991)

[10] Koufaris, M.: Applying the technology acceptance model and flow theory to online consumer behavior. Information Systems Research 13(2), 205–223 (2002)

[11] Taylor, S., Todd, P.A.: Understanding Information Technology Usage: A Test of Competing Models. Information Systems Research 6(2), 144–176 (1995)

[12] Rogers, E.M.: Diffusion of Innovations. The Free Press, New York (1995)

[13] Agarwal, R., Prasad, J.: A conceptual and operational definition of personal innovativeness in the domain of information technology. Information Systems Research 9(2), 204–215 (1998)

[14] Yi, M.Y., Jackson, J.D., Park, J.S., Probst, J.C.: Understanding information technology acceptance by individual professionals: Toward an integrative view. Information & Managemen 43(3), 350–363 (2006)

[15] China Internet Development Statistics Report 2007.7, China Internet Network Information Center (2007)

[16] Wu, J.H., Wang, S.C.: What drives mobile commerce An empirical evaluation of the revised technology acceptance model. Information & Management (42), 719–729 (2005)

[17] Joreskog, K.G., Sorbom, D.: LISREL8: Structural Equation Modeling with the SIMPLIS Command Language, Hove and London, NJ (1993)

[18] MacCallum, R.C., Browne, M.W., Sugawara, H.W.: Power analysis and determination of sample size for covariance structure modeling. Psychological Methods 1, 130–149 (1996)

[19] Bagozzi, R.P., Yi, Y.: On the evaluation of structural equation models. Journal of Academy of Marketing Science 16(1), 74–94 (1988)

Markov Tree Prediction on Web Cache Prefetching

Wenying Feng, Shushuang Man, and Gongzhu Hu

Abstract. As Web accesses increase exponentially in the past decade, it is funda-
mentally important for Web servers to be able to minimize the latency and respond
to users' requests very quickly. One commonly used strategy is to "predict" what
pages the user is likely to access in the near future so that the server can prefetch
these pages and store them in a cache on the local machine, a Web proxy or a Web
server. In this paper, we present an approach to effectively make page predictions
and cache prefetching using Markov tree. Our method builds a Markov tree from
a training data set that contains Web page access patterns of users, and make
predictions for new page requests by searching the Markov tree. These predicted
pages are prefetched from the server and stored in a cache, which is managed using
the Least Recently Used replacement policy. Algorithms are proposed to handle
different cases of cache prefetching. Simulation experiments were conducted with a
real world data of a Web access log from the Internet Traffic Achieve and the results
show the effectiveness of our algorithms.

Keywords: Web cache, Web server, cache prefetching, Least-Recent-Used, Markov
model, Markov tree.

Wenying Feng
Department of Computing & Information Systems and Department of Mathematics,
Trent University, Peterborough, Ontario, Canada K9J 7B8
e-mail: wfeng@trentu.ca

Shushuang Man
Department of Mathematics & Computer Science, Southwest Minnesota State University,
Marshall, MN 56258, USA
e-mail: mans@southwestmsu.edu

Gongzhu Hu
Department of Computer Science, Central Michigan University, Mt. Pleasant,
MI 48859, USA
e-mail: hu1g@cmich.edu

R. Lee, N. Ishii (Eds.): Software Engineering, Artificial Intelligence, SCI 209, pp. 105–120.
springerlink.com   © Springer-Verlag Berlin Heidelberg 2009

1 Introduction

An interesting topic on the area of Web applications is to predict future user activities. Various approaches such as data mining and machine learning have been proposed and studied. It is shown that a good prediction algorithm can increase user satisfaction by fast responses. To the service provider, it not only makes financially benefits from happier users, but also provides insights of user behaviors.

Assume that a Web server is stored a fixed number of Web pages. User access (visiting Web pages) is viewed as an ordered sequence of requests. One of the prediction techniques presented in [3] is the Markov Tree approach. The idea is to record the probabilities of the request sequences (*n-grams*) in a tree data structure. When a new request arrives, the tree is searched to find the page that is most likely to be requested next. The predicted request is then prefetched to *proxies* that sit between users and servers. This method was previously applied to represent past activities in [6, 10].

The Markov tree prediction can also be used as a prefetching scheme for web caching. A cache may be on the user's machine, together with the server, or separately stands between the user and the server. It can be viewed as a smaller storage place to temperately hold the most popular pages. When a user requests a web page, the request is first forwarded to the cache. If the page is found in cache (cache hit), it is sent to the user directly without access to the server. The server is accessed only when the page is not found in cache (cache miss). In this case, the page is sent to the user from the server but a copy of the page is also stored in the cache. Typically, the size of the cache is much smaller than the server. Therefore, when the cache is full and a new page is needed to store in it, a replacement policy is needed to decide which page should be replaced to optimal the hit rate.

Similar as the proxies prefetching, an efficient prefetching scheme help to reduce the server's workload and save network bandwidth. The matrix prefetching algorithm presented in [7, 8] was shown to significantly improve cache hit rate for certain user request patterns. Hence, the server works more efficiently and the overall system performance is enhanced.

Following the study in [3], this paper aims to investigate the benefits of the Markov tree prediction to a Web cache that is implemented with a Least Recently Used (LRU) replacement policy [2, 15]. The main focus of this paper is to present the prediction models and corresponding algorithms for Web access prediction to cache. Simulation experiments were conducted using a real world data set of Web log. A portion of the data was used as the training set and several combinations of the data were used as the testing sets for four caching schemes. Performance comparisons are discussed by cache hit rate, prediction accuracy, server access rates due to cache miss, and server access rate due to prediction. The results show that the proposed approach produced high cache hit rates and low server access rates that are essential to reduce the response latency.

The paper is organized as follows. After briefly reviewing the related work in Section 2, we describe the proposed Markov tree approach in Section 3. In Section 4, we show the algorithms for prefetching for several caching scenarios. Simulation experimental results are given in Section 5, and finally Section 6 concludes the paper with a summary and future work.

2 Related Work

Research on Web access prediction can be traced back to the time when the Web became available in the 90's. Early work on this topic include predictive prefetching [11] and use of path profiles for describing HTTP request behavior [13]. Other types of methodologies include applying machine leaning idea to caching system [7] and Markov models. A survey on Web prefetching performance metrics can be found in [5].

Markov models are designed to study stochastic processes, for which a Web access sequence is considered such a process. Traditional Markov models, however, have serious limitations as stated in [12]. Variations of Markov models have been developed to overcome these limitations. A hybrid-order tree-like Markov model (HTMM) was proposed in [16] that combines a tree-like Markov model method and a hybrid-order method. The paper [4] presented a selective Markov model that builds a model with reduced state complexity by intelligently selecting parts of different order Markov models. An adaptive weighting hybrid-order Markov model was described in [9] that optimizes HTMM by minimizing the number of nodes in HTMM and improve the prediction accuracy.

Markov tree [6, 14] is considered a more capable mechanism for representing past activity in a form usable for prediction. In a Markov tree, the transitions from the root node to its k-th level descendants represent the probabilities in a $k - 1$-th order Markov model. An algorithm for building a Markov tree was given in [3]. The work presented in this paper is based on [3] with modifications.

3 Markov Tree Creation and Search

Given a *sequence* (ordered actions such as Web requests), each item in the sequence is stored in a node of a Markov tree. The node contains the identifier (i.e. name) of the item and other information. A sample node is given in Fig. 1.

The item is recorded as X, has been encountered 5 times in this sequence (*selfCount*), has 2 children (*numChildren*), these children have been seen 4 times (*childrenCounts*), and a 0.05 probability from its parent to this node.

3.1 Building Markov Tree from a Training Set

A training set is divided into a number of sequences. Each sequence starts with a special page called starting page and ends before the next starting page. The

Fig. 1 A sample node in a
Markov tree

starting page is the default page for the targeting website. For example, it is
index.htm for the training set used in this paper. All of the sequences instead
of user sessions are then used to build a Markov Tree by adopting the algorithm
introduced in [3]. The tree is initialized to contain the root node only. A method
named createMarkovTree is then called for each sequence in a loop. A modified
pseudo code of the method is given in Listing 1 where s is a sequence containing n
pages $s[i], i = 0, 1, \cdots, n - 1$, and h represents the height of the tree.

Listing 1 Pseudo code to build Markov tree

```
for  level = 1 to min(n,h) do
  for  i = 0 to min(n−level,0) do
    parent ← root;
    if  level = 1 then
        increment  parent.selfCount;
    for  j=0 to level −1 do
        if  j = level −1 then
            increment  parent.childrenCounts;
        k ← 0;
        found ← false;
        while k < parent.childrenCounts −1 and not found do
            child ← parent.getChild[k];
            if s[j] = child page then
                found ← true;
        end
        if not found then
            increment  parent.numChildren;
            child ← new  CNode(s[j]);
            parent.add(child);
        parent ← child;
        if j = level −1 then
            increment  parent.selfCount;
    end
  end
end
```

After the tree is built, another method named calculatePropability is called
to calculate the probability for each node hence each branch or pattern. A sample
Markov tree of height 4 with a training set of 20 pages in 7 sequences is shown in
Fig. 2. Each node of the tree contains information about it's self-count, number of
children, children counts and the probability from it's parent node to the node itself.

3.2 Tree Searching for Prediction

Once the tree is built against a training set, the patterns of page requesting orders
with their probabilities are recorded in the tree. They are then used to predict a user's
next request in a testing set by identifying the pattern with the pages the user just

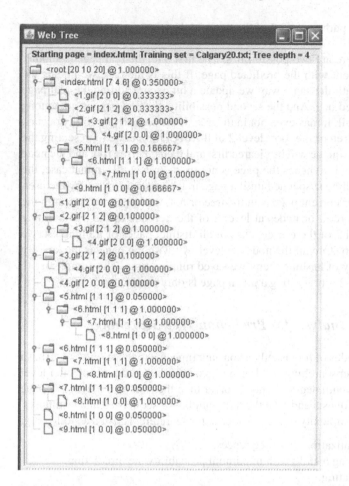

Fig. 2 Sample Markov tree of height 4 built with 20 pages in 7 sequences

browsed. We call the later a searching path whose length is dynamically determined up to a fixed small number, which is one less than the depth of the tree.

We use a fixed size array of Node as a queue (queue size = depth of the tree -1) to store the path from the root to current page, called searching path. For each sequence, which begins with the starting page, we initialize the searching path to contain the root of the tree and the starting page. To predict a user's next request, we compare the probabilities of the children of the most descendant node in the searching path. The page represented by the first highest probability child node is selected as the predicted page. It is then compared with the real request in the testing set. The prediction either hits or misses.

In case of hit, the predicted page will be added to the queue of searching path if it is not a leaf node, otherwise the queue will be cleared first before adding the new page to it. In other words, since a leaf is an indicator of the end of a branch, we redirect the searching path to level 2 of the tree and the queue contains root and the predicted node only.

In case of miss, there are two possibilities. The first one, the real page is another child of the same parent with the predicted page. If this is the case we update the queue of searching path the same way we update a hit except adding the real page instead of the predicted one. And the second possibility, the real page is not among the children nodes at all. It may even not in the tree. If the second possibility occurs, we search direct children of the root (level 2 of the tree) for a node representing the real page. If found, the queue will be cleared first and then have the root and the node added to it. If not found, it means the page is not in the tree at all. In that case, the prediction process will be suspended until a page in the tree is browsed by the user.

To monitor whether current page is in the tree or not, we only need to search the direct children of the root, or nodes at level 2 of the tree due to the tree building algorithm. Since level 2 of the tree consists of all distinct pages of training set, we can use a hash table to store all the nodes at level two with the page as the key. By scarifying the memory of hashing them, we saved running time. The complexity to search a node at level 2 representing a given page is only a constant.

3.3 Complexity Analysis for Prediction Tree Searching

Searching children nodes of root is only a constant time due to the adoption of a hash table. We only consider searching children nodes of non-leaf nodes at and after level 2. Let M be the maximum degree of nodes other than the root of the tree. Let N be the total number of requests and S be the total number of sequences in a testing set. The computational complexity of the worst case for searching is listed as following:

- $O(S)$ for path initialization of all sequences.
- $O(MN)$ for searching a child with maximum probability, i.e. prediction.
- $O(N)$ for path updating.
- Total: $O(S) + O(MN) + O(N) = O(MN)$, where $S \ll N$.

Since the degree of a non-root node is no more than the number of links contained in the web page represented by the node, it is safe to replace M by the maximum number of links contained in a single web page on a web server. It is a constant. So over all, the worst case complexity is $O(N)$. The searching used for prediction is linear with the help of a hash table. Otherwise, it would have been $O(SN)$, for $S \geq$ number of users.

4   Web Caching Prefetching by Markov Tree

Four algorithms are given in this section for handling the four cases of cache prefetching. Algorithm 1 performs predict prefetching on every request. Predict prefetching for cache miss and cache hit are given in Algorithm 2 and Algorithm 3, respectively. Algorithm 4 performs no prediction. The input to each of the four algorithm is the same that is a user's request for page p.

Algorithm 1. Request-predict-prefetching

Input: p: a page requested.

```
1  begin
2      if cache hit then
3          q ← next page prediction;
4          if q is not in cache then
5              Prefetching q from server;
6              Moved q to cache;
7              if cache is full then
8                  Apply LRU;
9                  Mark q the least recently used;
10     else
11         q ← next page prediction;
12         if q is in cache then
13             Fetch the missed page p;
14             Move p to cache (apply LRU if cache is full);
15         if q is not in cache (prefetching) then
16             Go to server to find p and q;
17             Move both pages to cache;
18             Forward p to the client;
19             p is marked most recently used;
20             q is marked least recently used;
21     end
22 end
```

Algorithm 2. Miss-predict-prefetching

Input: p: a page requested.

```
1  begin
2      if cache hit then
3          Forward p to client; // No prediction, no prefetching;
4      else
5          q ← next page prediction;
6          if q is in cache then
7              Fetch the missed page p;
8              Move p to cache (apply LRU if cache is full); // No prefetching;
9          if q is not in cache (prefetching) then
10             Go to server to find p and q;
11             Move both pages to cache;
12             Forward p to the client;
13             p is marked most recently used;
14             q is marked least recently used;
15     end
16 end
```

Algorithm 1 is for prediction and prefetching. When the user's request page p is in the cache, the algorithm makes a page prediction by searching the Markov tree as described in sub-Section 3.2. The predicted page q is put in the cache if it is not

---

Algorithm 3. Hit-predict-prefetching

   Input: p: a page requested.
1 **begin**
2   **if** *cache hit* **then**
3         $q \leftarrow$ next page prediction;
4   **if** *q is not in cache* **then**
5   Prefetching q from server;
6   Moved q to cache;
7   **if** *cache is full* **then**
8   Apply LRU;
9   Mark q the least recently used;
10   **else**
11   // no page prediction;
12   Go to server to find p;
13   Move p to cache (apply LRU if cache is full);
14   Forward p to the client;
15   p is marked most recently used;
16   **end**
17 **end**

Algorithm 4. No prediction or prefetching (normal cache)

   Input: p: a page requested.
1 **begin**
2   **if** *cache hit* **then**
3   Forward p to client;
4   **else**
5   Go to server to find p;
6   Move p to cache (apply LRU if cache is full);
7   Forward p to the client;
8   p is marked most recently used;
9   **end**
10 **end**

---

already in, and the LRU cache replacement is performed if cache is full. If p is not in the cache, a prediction q is made and prefetched if not in cache. p is fetched from the server and cached.

Explanations for Algorithm 2 – 4 are similar because they are "spacial cases" of Algorithm 1. We shall not explain further.

5 Simulation Results and Analysis

Simulation experiments were conducted to test our algorithms with the Markov tree structure. We shall briefly describe the data sets and report the results of four experiments that used the same training set but different testing sets.

5.1   Data Sets

The data for our experiments was the `Calgary-HTTP` data set from the Internet Traffic Archive [1]. The data set contains 786 distinguished URLs and 1477 non-trivial sequences with over 726,000 records of one-year HTTP logs from a CS departmental Web server. We used only 50,000 records in our experiments with the first 10,000 as the training set but four different testing sets:

1. Records 10004–30000 of the log file (excluding training set).
2. Records 10004–50000 of the log file (excluding training set).
3. Records 1–30000 of the log file (including training set).
4. Records 1–50000 of the log file (including training set).

For each of these testing data sets, the simulation was done for four schemes: always do prediction, predict only if a cache miss occurs, predict only if a cache hit occurs, and never predict, corresponding to the Algorithms 1 – 4.

The Markov tree built based on the training set has a depth 5. Tree searches for page predictions in our experiments went down to level 4 of the Markov tree.

5.2   Results for Cache Size 5%–30%

We used the cache size of 5% – 30% in these experiments. Cache size is measured as (*number of pages cached*) / (*total number of pages*).

The results for test case 1 (records 10004 – 30000 as the testing data) are shown in Table 1 that contains six subtables. Table 1(a) shows the cache hit rates. The prediction accuracy rates are given in Table 1(b) that is the ratio of the correct predictions vs. the total predictions. Server access rate is calculated as the ratio of server accesses vs. the total number of requests. Table 1(c) shows the total server access rates that is the sum of two cases: server access due to prediction in Table 1(d), and server access due to cache miss in Table 1(e). The total number of predictions made by the algorithms are given in Table 1(f). Note that Algorithm 4 makes no predictions, so no results using this algorithm are shown in the sub-tables involved in prediction.

One of the important performance measures for a web cache is its hit rate. Table 1(a) presents hit rates collected for four different scenarios. As expected, Algorithm 1 produces the highest hit rate with the sacrifice of highest server access rate shown in Table 1(c). The hit rate generated from Algorithm 2 is slightly less than that of Algorithm 1 but higher than that of Algorithm 3. However, server access rate of Algorithm 2 is the lowest of the first three algorithms. If the network traffic is more concerned, Algorithm 2 is a better prefetching mechanism than Algorithms 1 and 3, especially for small size caches.

As the cache size is increased, although all algorithms produce higher hit rates, Algorithm 4 has the biggest increase. For instance, when the cache size reaches 30%, differences by hit rate among the four approaches are much smaller; in fact, differences among algorithms 2, 3, 4 are less than 2%. On the other side, the rates of

Table 1 Testing set: ProcessedCalgory10004_To_30000.txt (excluding training set)

(a) Cache hit rate

Algorithm	Cache Size					
	5%	10%	15%	20%	25%	30%
1	0.723709	0.776794	0.815890	0.845175	0.861679	0.882510
2	0.649492	0.711935	0.755258	0.785096	0.81262	0.832394
3	0.591074	0.672134	0.724062	0.763762	0.804267	0.824595
4	0.569337	0.65231	0.707910	0.749874	0.785398	0.810003

(b) Rate of prediction success

Algorithm	Cache Size					
	5%	10%	15%	20%	25%	30%
1	0.479623	0.479623	0.479623	0.479623	0.479623	0.479623
2	0.449676	0.437137	0.436834	0.439806	0.438207	0.436284
3	0.500000	0.498769	0.493842	0.493238	0.492175	0.489842

(c) Total server access rate

Algorithm	Cache Size					
	5%	10%	15%	20%	25%	30%
1	0.737144	0.601439	0.523850	0.455721	0.421103	0.345778
2	0.511724	0.408624	0.337828	0.290329	0.249220	0.217772
3	0.594193	0.490037	0.451142	0.361779	0.334004	0.317852
4	0.430663	0.347690	0.292090	0.250126	0.214602	0.189997

(d) Server access rate due to prediction

Algorithm	Cache Size					
	5%	10%	15%	20%	25%	30%
1	0.460853	0.378233	0.339740	0.300896	0.282782	0.228288
2	0.161216	0.120560	0.093086	0.075425	0.061840	0.050166
3	0.185267	0.162172	0.175204	0.125541	0.138271	0.142447

(e) Server access rate due to cache miss

Algorithm	Cache Size					
	5%	10%	15%	20%	25%	30%
1	0.276291	0.223206	0.184110	0.154825	0.138321	0.117490
2	0.350508	0.288065	0.244742	0.214904	0.187380	0.167606
3	0.408926	0.327866	0.275938	0.236238	0.195733	0.175405
4	0.430663	0.347690	0.292090	0.250126	0.214602	0.189997

(f) Total predictions made

Algorithm	Cache Size					
	5%	10%	15%	20%	25%	30%
1	13250	13250	13250	13250	13250	13250
2	4481	3619	3095	2683	2298	2056
3	7810	8936	9663	10204	10799	11075

prediction success for Algorithms 1, 2, and 3 are not affected by cache size (Table 1(b)) but server access rates is greatly decreased for larger caches (Table 1(c)). For Algorithm 1, the number of predictions is a constant (Table 1(f)), reduce of the sever access is mainly by the less cache miss (Table 1(a)). For Algorithm 2, the increase of cache size insures less cache miss and therefore less number of predictions (Table 1(f)). On the other hand, the higher number of predictions for Algorithm 3 (Table 1(f)) is caused by its higher hit rate from the bigger cache.

Algorithm 1 has the lowest server assess rate due to cache miss (Table 1(e)) because it always performs prediction and prefetching so the number of cache misses would be minimized. On the contrary, Algorithm 4 has the highest such rate since it does not do prediction and prefetching so it would generate more cache misses. The gap between the two extremes reduces as the cache sizes increases.

In conclusion, cache prefetching systems are more efficient for smaller caches. Cache miss-prefetching outperforms cache hit-prefetcing. If network traffic is a big concern, miss-prefetching is a better choice than predict-prefetching on every client request. For bigger size caches, the benefit from prefetching may not be worth considering the implementation of the algorithm and the server access rate increased due to the prediction (Table 1(d)).

Experiments were also done on other three testing sets (stated in sub-section 5.1). Since the results are similar to test case 1, we put them in the Appendix. The results show that these algorithms all have reasonably good cache hit rates (about 70% –

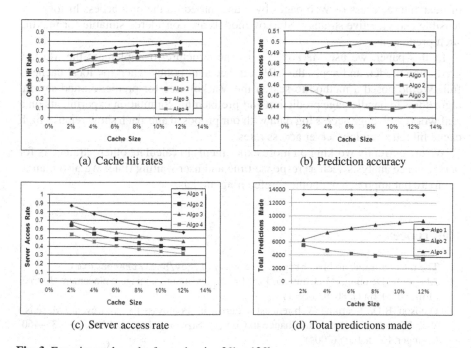

(a) Cache hit rates

(b) Prediction accuracy

(c) Server access rate

(d) Total predictions made

Fig. 3 Experimental results for cache size 2% – 12%

92%) when the cache size is at least 10%. Cache hit rates at these levels would reduce the page request and response latency in a significant way. The prediction accuracy stays pretty much stable regardless of the cache size.

5.3 Results for Cache Size 2%–12%

We also conducted simulations with refined cache sizes from 2% – 12% with 2% increment to calculate the cache hit rate, prediction accuracy, server access rate, and total predictions made. Rather than list the results in tables, we use charts for this group of experiments. The results are shown in Fig. 3(a) – Fig. 3(d).

These charts show that hit rates increase as the cache size is increased, but are somewhat lower than those with larger cache sizes of 10%–30% shown before. The rate of success prediction stay pretty much stable. The server access rate changes reversely proportional to the cache size. Algorithms 3 and 2 made similar number of predictions for 2% size cache but the difference increases as the size is increased.

6 Conclusion

Reducing Web access latency to improve the response time is a critical task as the Web becomes *the* vehicle for many users to retrieve information. Prediction of near-future access of Web pages by a user based on the past access history is a feasible and effective strategy. Markov models are considered suitable for making such predictions.

In this paper, we use Markov tree as the structure to implement the Markov prediction model, and apply the LRU replacement policy when the local cache is full. We proposed a method to search the Markov tree for prediction and several algorithms to handle page prediction and prefetching. Simulation experiments on a real world Web page access log data with our proposed approach show relative high cache hit rates and low server access rates.

We are currently working on more experiments to calculate other parameters for performance analysis such as response time and user waiting time. We also plan to compare our approach with other prefetching models.

References

1. The internet traffic archive, `http://ita.ee.lbl.gov/html/traces.html`
2. Aho, A.V., Denning, P.J., Ulmman, J.D.: Principles of optimal page replacement. Journal of the ACM 18(1), 80–93 (1971)
3. Davison, B.D.: Learning Web request patterns. In: Poulovassilis, A., Levene, M. (eds.) Web Dynamics: Adapting to Change in Content, Size, Topology and Use, pp. 435–460. Springer, Heidelberg (2004)
4. Deshpande, M., Karypis, G.: Selective Markov models for predicting Web-page access. ACM Transactions on Internet Technology 4(2), 163–184 (2004)

5. Doménech, J., Gil, J.A., Sahuquillo, J., Pont, A.: Web prefetching performance metrics: a survey. Performance Evaluation 63(9), 988–1004 (2006)
6. Fan, G., Xia, X.-G.: Maximum likelihood texture analysis and classification using wavelet-domain hidden Markov models. In: proceedings of the 34th Asilomar Conference on Signals, Systems, and Computers, Pacific Grove, CA, vol. 2, pp. 921–925 (2000)
7. Feng, W., Chen, H.: A matrix algorithm for Web cache prefetching. In: Proceedings of the 6th IEEE/ACIS International Conference on Computer and Information Science, Melbourne, Australia, July 11-13 2007, pp. 788–794. IEEE Computer Society Press, Los Alamitos (2007)
8. Feng, W., Vij, K.: Machine learning prediction and Web access modeling. In: Proceedings of the 31st Annual IEEE International Computer Software and Applications Conference, Beijing, China, July 23-27, 2007, pp. 607–612. IEEE Computer Society Press, Los Alamitos (2007)
9. He, S., Qin, Z., Chen, Y.: Web pre-fetching using adaptive weight hybrid-order markov model. In: Zhou, X., Su, S., Papazoglou, M.P., Orlowska, M.E., Jeffery, K. (eds.) WISE 2004. LNCS, vol. 3306, pp. 313–318. Springer, Heidelberg (2004)
10. Laird, P., Saul, R.: Discrete sequence prediction and its applications. Machine Learning 15(1), 43–68 (1994)
11. Padmanabhanand, V.N., Mogul, J.C.: Using predictive prefetching to improve World Wide Web latency. SIGCOMM Computer Communication Review 26(3), 22–36 (1996)
12. Popa, R., Levendovszky, T.: Markov models for Web access prediction. In: 8th International Symposium of Hungarian Researchers on Computational Intelligence and Informatics, pp. 539–550 (2007)
13. Schechter, S., Krishnan, M., Smith, M.D.: Using path profiles to predict HTTP requests. Computer Networks and ISDN Systems 30(1-7), 457–467 (1998)
14. Shafer, G., Shenoy, P.P., Mellouli, K.: Propagating belief functions in qualitative Markov trees. International Journal of Approximate Reasoning 1(4), 349–400 (1987)
15. Tanenbaum, A.S.: Modern Operating Systems, 2nd edn. Prentice-Hall, Englewood Cliffs (2001)
16. Xing, D., Shen, J.: A new Markov model for Web access prediction. Computing in Science and Engineering 4(6), 34–39 (2002)

Appendix – Experimental Results of More Testing Cases

This Appendix shows the results of three more testing cases with different testing data sets. Table 2 shows the results of the test case 2 (records 10004 – 50000). It has slightly higher cache hit rate, lower server access rate and made a lot more predictions.

Table 3 shows the results of test case 3 (records 1 – 30000) where the records in the training set is also included in testing. Comparing with test cast 1, this test case had higher cache hit and prediction success rates, lower server access rate, and much more predictions made.

The fourth test case is to use all the 50,000 records for testing including the training set. The results are shown in Table 4. This test case has the highest cache hit rate among the four test cases, and lowest server access rate.

Table 2 Testing set: ProcessedCalgory10004_To_50000.txt (excluding training set)

(a) Cache hit rate

Algorithm	Cache Size					
	5%	10%	15%	20%	25%	30%
1	0.741365	0.801389	0.841840	0.871098	0.891198	0.909436
2	0.682322	0.750094	0.795125	0.827325	0.856507	0.877739
3	0.636487	0.720208	0.777867	0.816734	0.851727	0.876682
4	0.618878	0.705768	0.762270	0.804936	0.837086	0.866972

(b) Rate of prediction success

Algorithm	Cache Size					
	5%	10%	15%	20%	25%	30%
1	0.466021	0.466021	0.466021	0.466021	0.466021	0.466021
2	0.428060	0.430224	0.425641	0.427613	0.421615	0.406786
3	0.482950	0.480867	0.479193	0.477336	0.474749	0.474968

(c) Total server access rate

Algorithm	Cache Size					
	5%	10%	15%	20%	25%	30%
1	0.653292	0.516113	0.424392	0.353752	0.304068	0.258233
2	0.445800	0.340922	0.272194	0.222611	0.181178	0.150386
3	0.536087	0.421952	0.357677	0.319967	0.265000	0.217127
4	0.381122	0.294232	0.237730	0.195064	0.162914	0.133028

(d) Server access rate due to prediction

Algorithm	Cache Size					
	5%	10%	15%	20%	25%	30%
1	0.394657	0.317501	0.266232	0.224850	0.195266	0.167669
2	0.128123	0.091017	0.067319	0.049936	0.037685	0.028125
3	0.172574	0.142160	0.135544	0.136701	0.116727	0.093809

(e) Server access rate due to cache miss

Algorithm	Cache Size					
	5%	10%	15%	20%	25%	30%
1	0.258635	0.198611	0.158160	0.128902	0.108802	0.090564
2	0.317678	0.249906	0.204875	0.172675	0.143493	0.122261
3	0.363513	0.279792	0.222133	0.183266	0.148273	0.123318
4	0.381122	0.294232	0.237730	0.195064	0.162914	0.133028

(f) Total predictions made

Algorithm	Cache Size					
	5%	10%	15%	20%	25%	30%
1	24780	24780	24780	24780	24780	24780
2	7256	5704	4680	3875	3183	2623
3	16070	18136	19585	20583	21445	22092

Table 3 Testing set: ProcessedCalgory30000.txt (including training set)

(a) Cache hit rate

Algorithm	Cache Size					
	5%	10%	15%	20%	25%	30%
1	0.777513	0.827731	0.861244	0.884101	0.904034	0.917916
2	0.684538	0.743092	0.782655	0.814891	0.839227	0.865479
3	0.626756	0.714420	0.767529	0.809244	0.844000	0.868672
4	0.590891	0.677546	0.731059	0.774286	0.811664	0.839092

(b) Rate of prediction success

Algorithm	Cache Size					
	5%	10%	15%	20%	25%	30%
1	0.549420	0.549420	0.549420	0.549420	0.549420	0.549420
2	0.553419	0.555001	0.565914	0.564807	0.563415	0.560423
3	0.539882	0.548173	0.546323	0.547872	0.548431	0.546341

(c) Total server access rate

Algorithm	Cache Size					
	5%	10%	15%	20%	25%	30%
1	0.684706	0.552437	0.465916	0.422958	0.342118	0.289244
2	0.468471	0.368706	0.303429	0.250487	0.212807	0.173311
3	0.569546	0.465647	0.408773	0.346958	0.295765	0.261143
4	0.409109	0.322454	0.268941	0.225714	0.188336	0.160908

(d) Server access rate due to prediction

Algorithm	Cache Size					
	5%	10%	15%	20%	25%	30%
1	0.462218	0.380168	0.327160	0.307059	0.246151	0.207160
2	0.153008	0.111798	0.086084	0.065378	0.052034	0.038790
3	0.196303	0.180067	0.176303	0.156202	0.139765	0.129815

(e) Server access rate due to cache miss

Algorithm	Cache Size					
	5%	10%	15%	20%	25%	30%
1	0.222487	0.172269	0.138756	0.115899	0.095966	0.082084
2	0.315462	0.256908	0.217345	0.185109	0.160773	0.134521
3	0.373244	0.285580	0.232471	0.190756	0.156000	0.131328
4	0.409109	0.322454	0.268941	0.225714	0.188336	0.160908

(f) Total predictions made

Algorithm	Cache Size					
	5%	10%	15%	20%	25%	30%
1	20609	20609	20609	20609	20609	20609
2	6449	5209	4430	3711	3209	2648
3	12725	14645	15813	16732	17499	18051

Table 4 Testing set: ProcessedCalgory50000.txt (including training set)

(a) Cache hit rate

Algorithm	Cache Size					
	5%	10%	15%	20%	25%	30%
1	0.773934	0.830878	0.866927	0.894976	0.911560	0.927237
2	0.702239	0.766014	0.810829	0.842243	0.872005	0.893485
3	0.652790	0.743406	0.801197	0.842324	0.875673	0.897939
4	0.627844	0.718198	0.775666	0.819393	0.856610	0.881053

(b) Rate of prediction success

Algorithm	Cache Size					
	5%	10%	15%	20%	25%	30%
1	0.513955	0.513955	0.513955	0.513955	0.513955	0.513955
2	0.507514	0.511492	0.512251	0.517447	0.513385	0.522442
3	0.511571	0.516500	0.516438	0.516733	0.515825	0.515448

(c) Total server access rate

Algorithm	Cache Size					
	5%	10%	15%	20%	25%	30%
1	0.628388	0.495295	0.401032	0.321740	0.264090	0.235799
2	0.425897	0.322728	0.252222	0.204304	0.160578	0.129728
3	0.527616	0.430955	0.361960	0.293792	0.231709	0.208254
4	0.372156	0.281802	0.224334	0.180607	0.143390	0.118947

(d) Server access rate due to prediction

Algorithm	Cache Size					
	5%	10%	15%	20%	25%	30%
1	0.402321	0.326173	0.267959	0.216717	0.175650	0.163036
2	0.128136	0.088742	0.063050	0.046547	0.032583	0.023213
3	0.180406	0.174361	0.163157	0.136115	0.107381	0.106192

(e) Server access rate due to cache miss

Algorithm	Cache Size					
	5%	10%	15%	20%	25%	30%
1	0.226066	0.169122	0.133073	0.105023	0.088440	0.072763
2	0.297761	0.233986	0.189171	0.157757	0.127995	0.106515
3	0.347210	0.256594	0.198803	0.157676	0.124327	0.102061
4	0.372156	0.281802	0.224334	0.180607	0.143390	0.118947

(f) Total predictions made

Algorithm	Cache Size					
	5%	10%	15%	20%	25%	30%
1	32139	32139	32139	32139	32139	32139
2	9050	7179	5714	4700	3773	3030
3	21088	24091	26007	27401	28500	29260

Frameworks for Web Usage Mining

Haeng-Kon Kim and Roger Y. Lee

Abstract. Web mining is a rapid growing research area. It consists of Web usage mining, Web structure mining, and Web content mining. Web usage mining refers to the discovery of user access patterns from Web usage logs. Web structure mining tries to discover useful knowledge from the structure of hyperlinks. Web content mining aims to extract/mine useful information or knowledge from web page contents. This tutorial focuses on Web Content Mining. A web browser of a limited size has difficulty in expressing on a screen information about goods like an Internet shopping mall. Page scrolling is used to overcome such a limitation in expression. For a web page using page scrolling, it is impossible to use click-stream based analysis in analyzing interest for each area by page scrolling. In this study, a web-using mining system is presented, designed, and implemented using page scrolling to track the position of the scroll bar and movements of the window cursor regularly within a window browser for real-time transfer to a mining server and to analyze user's interest by using information received from the analysis of the visual perception area of the web page.

Keywords: web mining, web usage, web information extract.

1 Introduction

Web content mining is related but different from data mining and text mining. It is related to data mining because many data mining techniques can be applied in Web content mining. It is related to text mining because much of the web contents are texts. However, it is also quite different from data

Haeng-Kon Kim
Department of Computer information & Communication Engineering,
Catholic Univ. of Daegu, Korea
e-mail: hangkon@cu.ac.kr

Roger Y. Lee
Software Engineering Information Technology Institute,
Central Michigan University, USA
e-mail: lee1ry@cmich.edu

R. Lee, N. Ishii (Eds.): Software Engineering, Artificial Intelligence, SCI 209, pp. 121–134.
springerlink.com   © Springer-Verlag Berlin Heidelberg 2009

mining because Web data are mainly semi-structured and/or unstructured, while data mining deals primarily with structured data. Web content mining is also different from text mining because of the semi-structure nature of the Web, while text mining focuses on unstructured texts. Web content mining thus requires creative applications of data mining and/or text mining techniques and also its own unique approaches. In the past few years, there was a rapid expansion of activities in the Web content mining area[1, 2]. This is not surprising because of the phenomenal growth of the Web contents and significant economic benefit of such mining. However, due to the heterogeneity and the lack of structure of Web data, automated discovery of targeted or unexpected knowledge information still present many challenging research problems. In this tutorial, we will examine the following important Web content mining problems and discuss existing techniques for solving these problems. Some other emerging problems will also be surveyed.

- **Data/information extraction:** Our focus will be on extraction of structured data from Web pages, such as products and search results. Extracting such data allows one to provide services. Two main types of techniques, machine learning and automatic extraction are covered.

- **Web information integration and schema matching:** Although the Web contains a huge amount of data, each web site (or even page) represents similar information differently. How to identify or match semantically similar data is a very important problem with many practical applications. Some existing techniques and problems are examined.

- **Opinion extraction from online sources:** There are many online opinion sources, e.g., customer reviews of products, forums, blogs and chat rooms. Mining opinions (especially consumer opinions) is of great importance for marketing intelligence and product benchmarking. We will introduce a few tasks and techniques to mine such sources.

- **Knowledge synthesis:** Concept hierarchies or ontology are useful in many applications. However, generating them manually is very time consuming. A few existing methods that explores the information redundancy of the Web will be presented. The main application is to synthesize and organize the pieces of information on the Web to give the user a coherent picture of the topic domain.

- **Segmenting Web pages and detecting noise:** In many Web applications, one only wants the main content of the Web page without advertisements, navigation links, copyright notices. Automatically segmenting Web page to extract the main content of the pages is interesting problem. A number of interesting techniques have been proposed in the past few years.

All these tasks present major research challenges and their solutions also have immediate real-life applications. The tutorial will start with a short motivation of the Web content mining. We then discuss the difference between web content mining and text mining, and between Web content mining and data mining. This is followed by presenting the above problems and current state-of-the-art techniques. Various examples will also be given to help participants to better understand how this technology can be deployed and to help businesses. All parts of the tutorial will have a mix of research and industry flavor, addressing seminal research concepts and looking at the technology from an industry angle. A lot of methods including log analysis based web data mining, eye tracking, and mouse tracking are being used to evaluate interest in web pages. A series of processes for collecting, accumulating, and analyzing data to evaluate interest require enormous capital in constructing a relevant system. Unfortunately, additional human resources and time have been necessary in collecting and accumulating data efficiently through the system constructed. In particular, there are many researches in web data mining based on log analysis, which provides convenience in information collection and includes usage information for all visitors. For an Internet shopping mall, it is difficult to express all information about goods through a web browser of a limited size.

To overcome such a limitation, page scrolling is used to identify it; in this case, however, the existing log based analysis may not be useful in analyzing interest in information a user wants.

Based on user interface environment, a web is not real space that users can feel directly with their hands but virtual one that they may feel indirectly through an input device such as a keyboard or a mouse. Therefore, it can be said that actual interaction between interface and a user occurs through an input device of PC.

In this paper, we present a web usage mining model using page scrolling to collect the position of the scroll bar of a web browser and movements of a window cursor regularly, transfer the results to the mining server in real time, and analyze the visual recognition area of the relevant web page and captured images and collected data through the mining server. This paper has the following construction. Relevant researches in Chapter 2 consider how to segment pages based on visual recognition and analyze the recognition rate by web page areas and web usage mining; Chapter 3 designs this web usage mining system using page scrolling. Chapter 4 implements the web usage mining system using page scrolling; Chapter 5 draws conclusions and presents subjects for a future study.

2 Related Work

2.1 Web Data

One of the key steps in Knowledge Discovery in Databases is to create a suitable target data set for the data mining tasks. In Web Mining, data can be collected at the server side, client-side, proxy servers, or obtained from an organization's database (which contains business data or consolidated Web data). Each type of data collection differs not only in terms of the location of the data source, but also the kinds of data available, the segment of population from which the data was collected, and its method of implementation. There are many kinds of data that can be used in Web Mining.

- Content: The real data in the Web pages, i.e. the data the Web page was designed to convey to the users. This usually consists of, but is not limited to, text and graphics.
- Structure: Data which describes the organization of the content. Intra page structure information includes the arrangement of various HTML or XML tags within a given page. This can be represented as a tree structure, where the tag becomes the root of the tree.
  The principal kind of inter-page structure information is hyper-links connecting one page to another
- Usage: Data that describes the pattern of usage of Web pages, such as IP addresses, page references, and the date and time of accesses.
- User Profile: Data that provides demographic information about users of the Web site. This includes registration data and customer profile information.

2.2 Web Usage Mining

Visitor search behaviors are recorded in a web server log file, along with information on user IP address, date and time of accessing a web page, how to request, URL of the page accessed, protocol being used in transferring data, status codes, the number of bytes transferred, and so on[4].

Figure 1 shows web mining processing for user behaviors using log files.

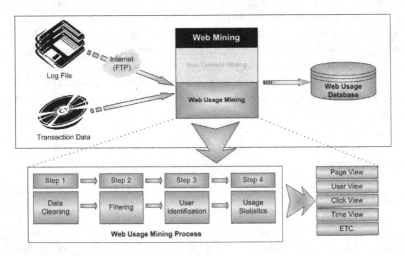

Fig. 1 Web Usage Mining Process

In a data cleaning process, information on visitor behaviors recorded in a log file was parsed to extract information such as IP address, time of connection, and re-quested pages. Filtering aims to remove information unnecessary for visitor search behaviors, such as image, by using the information cleaned.

User identification aims to track accurate information on visits by using information about a visitor's IP address, the amount of time for maintaining session, and the browser being used and that about log-in. Statistical visitor information, such as statistical visit date and time, the number of visits, the revisit rate, the browser being used, and OS being used, was extracted and analyzed from multiple viewpoints of page, user, click-stream, and time series[2].

2.3 Recognition of Common Areas in a Web page Using Visual Information

Figure 2 shows the areas segmented by Milos Kovacevic to designate an interest area and analyze the recognition rate by web page areas: H (Header) for 200 pixels at the top of the web page, F (Footer) for 150 pixels at its bottom, LM (Left Menu) for 15 percent on the left, RM (Right Menu) for 15 percent on the right, and C (Center) for the remaining area[3].

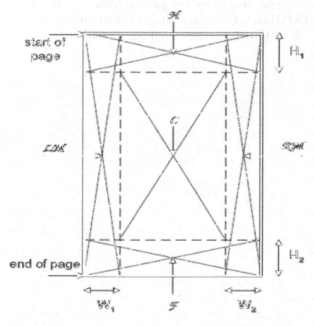

Fig. 2 Position of areas of interest in a page

Table 1 RECOGNITION RATE BY INTEREST AREAS

	H	F	LM	RM	Overall
Not recognized	25	13	6	5	3
Bad	16	17	15	14	24
Good	10	15	3	2	50
Excellent	49	55	76	79	23

Table 1 shows the results from the test Milos Kovacevic implemented for the recognition rate by interest areas: 59 percent for H, 70 percent for F, 79 percent for LM, and 81 percent for RM. Thus, the recognition rate was low at the top and bottom of the web page and there was no significant difference in the rate between LM and RM.

2.4 Vision-Based Page Segmentation

People use a web browser to explore a web page, which is provided in two-dimensional expression of many visual blocks segmented by lines, blanks, image, color, and so on. As a page segmentation method based on visual recognition, which was presented by Deng C., VIPS simulates how people understand web page layout through their visual recognition capacity, and uses a document object model (DOM) structure and visual information repetitively to extract visual blocks, ultimately detecting visual segmentation elements,

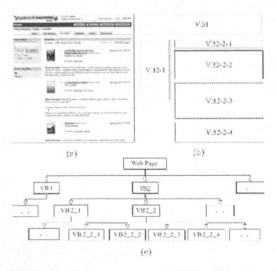

Fig. 3 The layout structure and vision-based content structure of an example page

constructing a content structure, and thus extracting a content structure based on visual recognition[5], [8].

Figure 3 shows a content structure based on visual recognition for a sample page, detecting visual blocks as in figure 3(b) and expressing the content structure as in figure 3(c), thus reflecting the semantic structure of the page[5].

3 Web Usage Mining System Using Page Scroll

3.1 *Web Usage Mining System Using Page Scroll*

A web site visually provides various multimedia contents (web contents), such as text, image, audio, and moving pictures, which are inserted in a web page as in figure 4. The monitor being most frequently used now has the resolution of 1024×768 pixels; by using this monitor to execute a web browser, the maximum 995×606 pixel web page can be expressed on the web browser. In figure 4, a 788×1375 pixel web page is actually shown; here, the browser, which is 995 pixels in width, is enough to express 788 pixel information, thus showing no horizontal scroll bar but a vertical scroll bar is shown to express

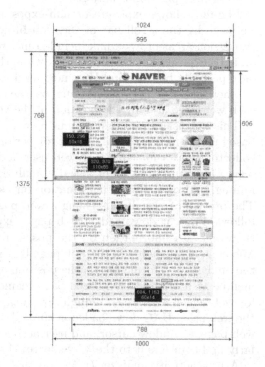

Fig. 4 Exposure of web page

Fig. 5 Exposure or non-exposure of information on a web page

the remaining information as it is 606 pixels in length and not enough to express 1375 pixel information[8].

Figure 5 shows exposure or non-exposure of information provided on a webpage by the scroll bar of the web browser; with HA, HB, and HC for expression areas of hyperlink included in a web page, the first scene had HA and HB exposed but HC unexposed while the second scene had HB exposed only partially but HA and HC unexposed.

Therefore, it is necessary to analyze the exposure and recognition rate for each web page area by the position of the vertical scroll bar.

This web usage mining system using page scrolling is composed of a vision-based usage mining (VUM) server for performing mining and a user activity collector (UAC) for collecting web user activities as in figure 6. The VUM server has the functions of visually analyzing a web page, or the target of mining, and of analyzing information on user activities collected via UAC. The process of visually analyzing a web page, or the target of mining, is as follows. First, it downloads a web page, expresses it visually in DOM, and then analyzes information on the position and size of visual expression of web contents included in the web page by the DOM analyzer in pixels. With the analysis of information on the position and size, it generates an area segmented visually by vision-based page segmentation (VPS), generates image for web screen and areas through screen capture, and then inserts UAC into the web page. UAC downloaded in the web browser collects information on a web page, a window cursor, and user activities including page scrolling regularly and transfers it to the VUM server through a communication module.

A user activity analyzer (UAA) in the VUM server analyzes information on web page log and user activities and stores it in database.

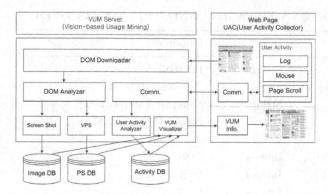

Fig. 6 Web Usage Mining System Using Page Scroll

The VUM visualizer visualizes information on visual analysis and user activities and provides it in the web page mode.

3.2 Usage Information Collection and Analysis

This usage information collection and analysis system cannot operate by the existing usage information collection method, or web server log file analysis, or by the log information collection method using java script, or simple click-stream. In addition to information on click-stream, it is therefore necessary to collect information on user behaviors such as the position of the window cursor and that of the scroll bar of the web browser and send it to the analysis system for analysis. Figure 7 shows the work flow for collecting and analyzing user information.

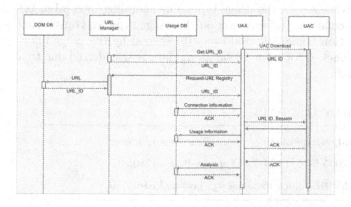

Fig. 7 Work-flow for collecting and analyzing user information

Table 2 INFORMATION ON WEB PAGE CONNECTION TO TRANSFER

	Type	Example
IP	String	127.0.0.1
URL	String	/Interest/49.html
Referrer URL	String	/Interest/Interest.htm
Time	Date	2005-6-15/13-57-7.858
Time Zone	Integer	-9
Screen Resolution(Width)	Integer	1280
Screen Resolution(Heigth)	Integer	1024
Browser Size(Width)	Integer	1272
Browser Size(Heigth)	Integer	915
Colors	Integer	16
Etc.	String	cookie=y,java=y,history=0

User information collection consists of two stages according to the properties of the web.

At the first stage, when a user connects a web page, the user information collection module inserted in the web page is executed, consequently sending information on web page connection and collecting that on the usage type. After receiving information on web page connection, the usage information analyzer analyzes the information, stores it in usage database, and assigns URL ID and a session to it. Table 2 shows information on web page connection to transfer; Table 3 shows log data transferred. At the second stage, information on the usage type is collected and transferred at the designated sampling interval. After receiving information on the usage type, the usage information analyzer analyzes the information and stores it in the usage database.

Table 4 shows the collected information on the usage type; Table 5 shows the transferred information on the usage type. The sampling interval of 0.1 second was used in collecting information on the usage type. In transferring usage type informa-tion being collected every 0.1 second by the usage information analyzer, a package of 10 or 20 sampling data is transferred due to a great amount of traffic.

Table 3 EXAMPLE OF LOG DATA

url=/Interest/49.html,referrer=http://www.eyerank.com/Interest /Interest.htm,time=2005-6-15/13-57-7.858,zone=-9,sw=1280, sh=1024,bw=1272,bh=915,color=16,cookie=y, java=y,history=0

Table 4 USAGE DATA TYPE

	Type	ex.
URL ID	Integer	1
Sampling ID	Integer	1
Scroll Position of Browser	Integer	240
Cursor Position of Window(X)	Integer	449
Cursor Position of Window(Y)	Integer	340

Table 5 EXAMPLE OF USAGE TYPE INFORMATION

url=0
MT=T1S0X449Y340T2S0X449Y340T3S0X449Y340T4S0X44
9Y340T5S0 X449Y340T6S0X449Y340T7S0X449Y340T8S0X449Y340T9S0X449
Y340T10S0X449Y340

4 Implementation

This web usage mining system using page scrolling used 0.1 second, or a half of the optimum interval of 0.2 to 0.3 second being used by L. Granka's eye tracking analy-sis[6], as the interval for collecting information on user activities, and its implementation environment is as shown in the following table 6.

In this study, we designed and implemented a system to analyze visual recognition areas of a web page and to collect the position of the scroll bar of the web browser and that of the window cursor and a mining system to analyze the results. Figure 8 shows the results of collecting and analyzing 957,103 pieces of user behavior information on 22,199 page views for a user who connected a web page. Exposure time was analyzed in seconds per vertical pixel, with the maximum exposure area around 391 pixels, which was exposed for

Table 6 H/W ENVIRONMENT

	OS	CPU	RAM	HDD	
H/W #1	Windows 2003 Server	P-IV 3.6GHz x 2	6GB	74GBx4 (RAID 0+1)	DB-Server (MySQL 4.0)
H/W #2	Windows 2003 Server	P-IV 3.6GHz x 2	6GB	74GBx2	Web-Server (IIS 6.0)
H/W #3	Windows XP Professional	P-IV 3.0GHz x 1	1.5GB	120GB	Client

Fig. 8 Exposure time
and Click-Through

Fig. 9 Interest on visually segmented areas

21,717 seconds; the average click rate was analyzed by classifying the Y position, among hyperlink exposure positions, in 100 pixels. The exposure time
for an area included in the first scene of the web browser is relatively longer
than that for an area included in the second scene; the closer to the bottom,
the shorter exposure time; and the click rate is also proportional to exposure
time.

And the analysis of information on user activities through page scrolling
makes it possible to make detailed exposure analysis on the web page; it is
therefore easy to determine what area of the page is most exposed to a user.
Figure 9 shows the results from the analysis of interest through the analysis
of exposure time and that of window cursor activities on visually segmented
areas.

5 Conclusion

In this study, we designed and implemented a web usage mining system using page scrolling to collect the position of the scroll bar of a web browser and movements of a window cursor regularly, transfer the results to the mining server in real time, and analyze the visual recognition area of the relevant web page and captured images and collected data through the mining server. Many existing data collection and analysis methods are based on frequency using page view, Hits, algorithms, and so on. Based on simple click events that occur in a web browser, these methods have a limitation of analyzing records on a user's information reference.

To overcome such a limitation for the existing analysis methods, this system discovered user activities within a web browser, used the position of page scrolling and that of the cursor to measure interest in a web page, used information on scrolling in a long page and that on window cursor coordinates not used for web usage mining, and thus could analyze the user's interest in the web page accurately. The web usage mining system using page scrolling should be applied to Internet shopping malls to standardize the techniques of analyzing interest in goods, along with further researches in a web-based business process.

Acknowledgments. This work was supported by the Korea Science and Engineering Foundation (KOSEF) grant funded by the Korea government (MEST) (No. R01-2008-000-20607-0).

References

1. Yang, T., Zhang, H.: HTML Page Analysis Based on Vusual Cues. In: 6th International Conf. Document Analysis and Recognition (ICDAR) (2001)
2. Lee, C.D., Choi, B.J., Gang, Z., Kim, I., Park, K.S.: A Study on Detection Technique of Bookmarking using Favicon in Web Browser. In: Proc. International Conference on East-Asian Language Processing and Interenet Information Technology 2002, EALPIIT 2002, Korea Information Processing Society, pp. 427–433 (2002)
3. Kovacevic, M., Diligenti, M., Gori, M., Maggini, M., Milutinovic, V.: Reconition of Common Areas in a Web page Using Visual Information: a possible application in a page classification. In: Proc. the 2002 IEEE International Conference on Data Mining (ICDM), pp. 250–257 (2002)
4. Jung, Y.-H., Kim, I., Park, K.-S.: Design and Implementation of an Interestingness Analysis System for Web Personalization & Customization. Journal of Korea Multimedia Society 6(4), 707–713 (2003)
5. Cai, D., Yu, S., Wen, J.-R., Ma, W.-Y.: VIPS: a vision-based page segmentation algorithm, Microsoft Technical Report, MSR_TR-2003-79 (2003)

6. Granka, L., Joachims, T., Gay, G.: Eye-Tracking Analysis of User Behavior in WWW-Search, Poster Abstract. In: Proc. the Conference on Research and Development in Information Retrieval (SIGIR) (2004)
7. Choi, B.J., Kim, I., Shin, Y.W., Kim, K.H., Park, K.S.: A Study of Link-Competition in a Hyperlinked Environment. In: International Conf. Internet Computing, pp. 339–344 (2004)
8. Kim, I., Park, K.: Vision Based Web Usage Mining. In: Proc. The 2005 International Symposium on Multimedia Applications in Edutainment (MAEDU), pp. 18–21 (2005)
9. Pitkow, J., Pirolli, P.: Mining longest repeating subsequences to Predict WWW Surfing. In: Proceedings of the 1999 USENIX Annual Technical Conference (1999)
10. Pohle, C., Spiliopoulou, M.: Building and Exploiting Ad Hoc Concept Hierarchies for Web Log Analysis. In: Kambayashi, Y., Winiwarter, W., Arikawa, M. (eds.) DaWaK 2002. LNCS, vol. 2454, pp. 83–93. Springer, Heidelberg (2002)
11. Padmanabhan, B., Zheng, Z., Kimbrough, S.O.: Personalization from incomplete data: What you don't know can hurt. In: Proceedings of ACM SIGKDD International Conference on Knowledge Discovery and Data Mining, San Francisco, CA, pp. 154–163 (2001)
12. Srikant, R., Agrawal, R.: Mining Generalized Association Rules. In: Proceedings of the 21st International Conference on Very Large Databases, Zurich, Switzerland, pp. 407–419 (September 1995)

A Concept Semantic Similarity Algorithm Based on Bayesian Estimation

Wu Kui, Guo Ling, Zhou Xianzhong, and Wang Jianyu

Abstract. Traditional algorithms for semantic similarity computation fall into three categories: distance-based, feature contrast and information-based methods. The former two methods ignore the objective statistics, while the last one cannot obtain enough domain data. In this paper, a new method for similarity computation based on Bayesian Estimation is proposed. First, the concept encountering probability is assumed to be a random variable with a priori Beta distribution. Second, its priori parameters are designated by the distance-based similarity algorithm. And the posteriori encountering probability is calculated based on Bayesian Estimation. Thereby, the semantic similarity integrating the subjective experience with the objective statistic is acquired based on information-based method. Finally, our method is implemented and the performance is analyzed by WordNet.

1 Introduction

Ontology is playing a more and more important role in the effective organization, share and reuse of domain knowledge for software engineering, artificial intelligence, information retrieval, web services and other areas [1]. The computation of concept semantic similarity, as a foundation problem, determines the performances of these applications. Similarity is usually supposed to be a value that embodies how similar two concepts are in both subjective experience and objective factors. Subjectively, a unanimous similarity between two concepts cannot be given because different people understand things in different ways. But undoubtedly the similarity between a concept and its sub-concept should be greater than that between its super concept and sub-concept. For example, the similarity between "mammal" and "man" cannot be given, but it should be greater than the similarity between "animal" and "man" and less than that between "mammal" and "person". Objectively, each concept has the characteristic of its ancestor concept. So the similarity between two concepts is related to the commonality inheriting from their common ancestors. The more commonality they share, the more similar they are.

To compute the similarity between two concepts, many measures have been proposed, such as the distance-based algorithms [2,3,4], the feature contrast model [5]

Wu Kui, Guo Ling, Zhou Xianzhong, and Wang Jianyu
School of Automation, Nanjing University of Science & Technology

R. Lee, N. Ishii (Eds.): Software Engineering, Artificial Intelligence, SCI 209, pp. 135–144.
springerlink.com   © Springer-Verlag Berlin Heidelberg 2009

and the information content methods [6,7,8]. The distance-based algorithms are simple and intuitive, but the conceptual hierarchical network that the computation relies on has to be established manually, and each edge's weight in this network is also manually maintained, which cannot guarantee the objectivity of the similarity. The feature contrast model is similar to the way people understand things, but impractical because of the requirement of detailed and comprehensive description about objects. The information content methods, theoretically more persuasive, measures the similarity by the ratio between the amount of information needed to state the commonality of A and B and the information needed to fully describe what A and B are [8], but the large and authoritative samples in domain are very difficult to obtain. In this paper, a new semantic similarity approach called Bayesian Estimation Similarity (BES) is proposed.

The remainder of this paper is organized as follows. Basic principles are present in the next section. Our algorithm for similarity computation is described in Section 3 and its performance is evaluated and compared with six other algorithms in Section 4. Finally, the paper is concluded in Section 5.

2 Basic Principles

2.1 Definitions

Ontology is a theory about the existence which originates from philosophy domain. In 1993, Tom Gruber gave the first formal definition in information science [9]: An ontology is an explicit specification of a conceptualization. Various formal specifications about ontology were presented after that. Here we introduce the one given by W3C [10].

Definition 2.1. Ontology is expressed as $O = (C, R, I, A)$, in which C is a concept set, R a relation set, I an individual set, and A an axiom set.

Definition 2.2. Concept, also known as Class, refers to the name of an entity set in which each element has similar characteristics.

Definition 2.3. If concept A has the characteristic of concept B, we say concept A is a sub-concept of concept B, denoted by $A \sqsubseteq B$. The relation $\sqsubseteq$ is called subclass relation or inheritance. If $A \sqsubseteq B$ and no concepts satisfies the inheritance relation between A and B, we say concept A is a direct sub-concept of B, denoted by $A = B^{\sqsupset}$. The sub-concepts set of concept B is denoted by $ch(B)$. The number of direct sub-concepts of concept B is called the out-degree of concept B, denoted by $\deg^{\sqsupset}(B)$. Also the direct super concept of B, denoted by $B^{\sqsubset}$, and the super concept set of B, denoted by $an(B)$, can be defined as above.

Definition 2.4. The concept set in domain ontology constitutes a hierarchical lattice by inheritance relation. The top node of the lattice is called top concept, denoted by $\top$. The minimum number of edges between concept C and the top concept $\top$ is called the depth of concept C, denoted by $dep(C)$.

Definition 2.5. In domain ontology, let each concept in ontology be augmented with a function $P:C \to [0,1]$, such that for any $C \in C$, $P(C)$ is the probability of encountering an individual of concept C. Statistically, the probability can be taken simply as relative frequency $P(C)=freq(C)/N$, where $freq(C) = \sum_{c \in C} count(c)$ is the number of encountering an individual of concept C, and N is the total number of individuals observed in domain ontology.

Definition 2.6. In domain ontology, the similarity between concept   and   is calculated as follows [8]:

$$Sim(C_1,C_2) = \frac{2\log P(LCS(C_1,C_2))}{\log P(C_1) + \log P(C_2)} \tag{1}$$

where $LCS(C_1,C_2)$ denotes the least common super concept of concept C_1 and C_2.

2.2 Some Intuitions in Similarity

From formula (1), we can see that concept encountering probability is essential in similarity computation, while it relies on large and authoritative samples of a concerned domain. Here we present a new approach based on Bayesian Estimation to reduce this dependence. Before further description, we propose two basic assumptions: (1) in domain ontology, the proposition 'a is an individual of concept C' has two truth values: True or False. That is to say, the random variable $x := a \in C$ obeys 0-1 distribution; (2) If a is an individual of concept C, it must be an individual of the super concept of C. Therefore the encountering probability of a concept must be greater than that of its sub-concept and less than that of its super concept. Obviously, these two basic assumptions are acceptable. Based on them, we clarify some intuitions [6, 8, 11] about similarity as follows:

1)   When two concepts are identical, the similarity reaches a constant maximum, assumed to be 1, and it reaches 0 when no commonality exists between them, that is $0 \le Sim(C_1,C_2) \le 1$, and $Sim(C,C)=1$.

2)   The similarity between a concept and its sub-concept is related with the depth where they locate in the conceptual hierarchical lattice. The greater the depth, the smaller the difference between them, and the greater their semantic similarity. Specially, the formula $Sim(C^\sqsubset,C) \le Sim(C,C^\sqsupset)$ is valid.

3)   The similarity of a concept with its sub-concept is related with its out-degree.

Bayesian Statistics [12] is one of the most important branches in statistics. Although this approach has aroused a lot of controversies, such as an assumed prior distribution maybe not reasonable for a real problem, these arguments are not the mainstream in recent years. The majority of researchers have admitted that the Bayesian approach can work well in nearly all branches of statistics. In this paper, we use it to calculate concept encountering probability. This approach is summarized as follows: First, the concept encountering probability is assumed to be a random variable with a prior Beta distribution, since Beta density function can take on different shapes through parameter adjustment. Second, the distance-based similarity between a concept and top concept is designated as the priori parameters of concept encountering probability. And the posteriori encountering probability is calculated based on Bayesian Estimation. Last, the semantic similarity integrating the subjective experience with the objective statistic is acquired based on formula (1).

3 Concept Encountering Probability Computation

3.1 Encountering Probability Designation

Let a_i denote an individual in domain ontology and random variable $x := a_i \in C$ obey 0-1 distribution with the encountering probability P of concept C. We designate it with the distance-based similarity, which is related to the weighted sum of shortest path in concept hierarchical lattice. The weight of each edge is computed as follows:

$$w_C = \frac{1}{\rho^{\text{dep}(C)}} \cdot \frac{1}{\deg^{\beth}(C^{\beth})} \tag{2}$$

where the parameter $\rho > 1$ determines how semantic distance varies with concept depth in concept hierarchical lattice. Then the semantic distance between two concepts is:

$$D|_{C_1}^{C_2} = \text{Distance}(C_1, C_2) = \sum_{\min\{C_1; C_2\}} w_{C_i} \tag{3}$$

where $\min\{C_1, C_2\}$ denotes the node set of the shortest path between concept C_1 and C_2. So their similarity is:

$$S|_{C_1}^{C_2} = \text{Sim}(C_1, C_2) = \frac{\mu}{D + \mu} \tag{4}$$

in which the parameter μ is an subjective parameter. From this definition, we can see that for any concept C, $0 < \text{Sim}(\top, ch(C)) < \text{Sim}(\top, C) < \text{Sim}(\top, an(C)) \leq 1$ is valid.

Suppose that the concept similarity in formula (4) can be mapped to the encountering probability by a monotonic increasing function $\overline{SP}(\cdot)$, Let $\hat{P} = \overline{SP}(S|_C^\top)$,

$\hat{P}_M = \max\{\overline{SP}(S \mid_{C^\supset}^{\top})\}$, and $\hat{P}_N = \min\{\overline{SP}(S \mid_{C^C}^{\top})\}$, then concept encountering probability P is assumed to be a beta distribution with mode $\hat{P}$ and interval $[\hat{P}_M, \hat{P}_N]$.

3.2 The Calculation of Shape Parameters

Considering the standard form of Beta Distribution, we perform a linear transformation [13]:

$$\text{Beta}(\theta; \alpha, \beta) = \begin{cases} \dfrac{\theta^{\alpha-1} \times (1-\theta)^{\beta-1}}{B(\alpha, \beta)} & 0 \le \theta \le 1 \\ 0 & \text{others} \end{cases} \tag{5}$$

in which

$$\theta = (P - \hat{P}_M) / (\hat{P}_N - \hat{P}_M) \tag{6}$$

So its shape parameters are:

$$\alpha = (\frac{E - \hat{P}_M}{\hat{P}_N - \hat{P}_M})(\frac{(E - \hat{P}_M)(\hat{P}_N - E)}{D} - 1) \tag{7}$$

$$\beta = (\frac{\hat{P}_N - E}{E - \hat{P}_M})\alpha \tag{8}$$

where its expectation and variance are calculated as follows: $E = (\hat{P}_M + 4\hat{P} + \hat{P}_N) / 6$ and $D = (\hat{P}_N - \hat{P}_M)^2 / 36$.

3.3 The Posteriori of Encountering Probability

According to statistics, the frequency of concept C is k, and the total number of individuals observed is N. Then we calculate the estimation of θ to minimize Bayesian risk.

Let $L(\theta, \delta)$ denote lose function and $\delta(X)$ denote discrimination function. So the Bayesian risk is:

$$\begin{aligned} R_\pi(\delta) &= \int_\theta R(\theta, \delta)\pi(\theta)\mathrm{d}\theta \\ &= \int_\theta \int_X L(\theta, \delta(X))\pi(\theta)f(X, \theta)\mathrm{d}\mu(x)\mathrm{d}\theta \end{aligned} \tag{9}$$

in which $f(X, \theta)$ denotes the joint distribution of X and θ. According to Bayesian's Theorem, we can get:

$$\pi(\theta \mid x) = \frac{\pi(\theta)f(X, \theta)}{\int \pi(\theta)f(X, \theta)\mathrm{d}\theta} = c(X)\pi(\theta)f(X, \theta) \tag{10}$$

where $c(X)$ is a unitary constant. Substitute formula (10) into formula (9) and exchange integration order, then we have:

$$R_\pi(\delta) = \int_x [\int_\theta L(\theta, \delta(X))\pi(\theta \mid x) d\theta] c(X) d\mu(x)$$

Let

$$R_\pi(\delta \mid x) = \int_\theta L(\theta, \delta(X))\pi(\theta \mid x) d\theta \qquad (11)$$

So if a $\delta(X)$ minimizes $R_\pi(\delta \mid x)$, it also minimizes the Bayesian risk $R_\pi(\delta)$. Let the lose function $L(\theta, \delta(X)) = (\delta(X) - \theta)^2$, and substitute it into formula (11),

$$R_\pi(\delta \mid x) = \int_\theta (\delta(X) - \theta)^2 \pi(\theta \mid x) d\theta = a\delta^2(X) + b\delta(X) + c \qquad (12)$$

Where $a = \int_\theta \pi(\theta \mid x) d\theta$ $b = -2 \int_\theta \theta\pi(\theta \mid x) d\theta$ $c = \int_\theta \theta^2 \pi(\theta \mid x) d\theta$. So $\delta = -b/2a$ minimizes the posteriori risk $R_\pi(\delta \mid x)$. According to formula (10), we have:

$$\pi(\theta \mid x) = \frac{c(x)}{B(\alpha, \beta)} \cdot \theta^{k+\alpha-1} \times (1-\theta)^{n-k+\beta-1} \qquad (13)$$

Therefore, $\pi(\theta \mid x)$ also obeys Beta distribution, and its hyperparameters are $k+\alpha$ and $n-k+\beta$, that is:

$$\pi(\theta \mid x) = \frac{t^{k+\alpha-1} \times (1-t)^{n-k+\beta-1}}{B(k+\alpha, n-k+\beta)} \qquad (14)$$

So

$$\delta = -\frac{b}{2a} = \frac{\alpha+k}{\alpha+\beta+n} \qquad (15)$$

According to the linear transform of formula (6), we can get

$$P = \hat{P}_M + \delta(\hat{P}_N - \hat{P}_M) \qquad (16)$$

Thus, we get the posteriori encountering probability P based on the Bayesian Estimation. From formula (16), we can see that concept encountering probability P is an empirical revision based on statistics. With large samples, δ tends to be k/n, which emphasizes statistical features, while with small samples, especially $k=0$, it does not mean that the individuals of concept cannot be observed. Instead, it can be calculated by subjective experience.

4   The Implement and Evaluation

Based on the BES algorithm principle, we implement an application to compare our method with others. In this program, we analyze the similarity in computer domain which takes the WordNet[14] as data source. Part of a concept hierarchical structure is shown in figure 1:

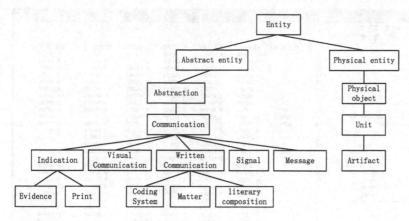

Fig. 1 The part of concept hierarchical structure

According to the structure, the priori parameters and posteriori encountering probability of each node are computed respectively. Because these computations are mathematical operation, the algorithm complexity is decided by the traversal strategy which is proved to be $O(n)$ [15]. The algorithm is described follows:

> **Input:** The concepts in conceptual hierarchical structure T
> **Output:** The posteriori encountering probability of each concept
> **Step 1:** Let top concept's $S = \hat{P} = \hat{P}_M = \hat{P}_N = 1$
> **Step 2:** Get the next uncalculated sub-concept N of top concept, Call *CalcValues*(N) to compute its encountering probability
> **Step 3:** If all the sub-concepts of the root is handled, the algorithm stops; otherwise, go to Step 2

The steps of *CalcValues*(*node N*) are as follows:

> **Input:** Concept N
> **Output:** The posteriori encountering probability of N
> **Step 1:** Calculate S and $\hat{P}$ of N
> **Step 2:** Travel each sub-concept of N, call *CalcValues*(N) recursively.
> **Step 3:** Let $\hat{P}_M$ of N equal to the maximal $\hat{P}$ of N's sub-concept, and $\hat{P}_N$ equal to the minimal $\hat{P}$ of N's super-concept.
> **Step 4:** According to formula (7), (8), (15), (16), calculate the posteriori encountering probability

The interface of our experiment program is shown in figure 2. In our experiments, let $\rho = 1.1$, $\mu = 0.5$, $\overline{SP}(x) = x$ and then randomly select 10 concept-pairs from the tree, calculate the similarity of each concept-pair, and compare these values with other methods in WordNet::Similarity [16]. The results are shown in Table1, where BES, HSO, JCN, WUP, LIN, RES and LCH stand for ours, Hirst & St-Onge's [2], Leacock & Chodorow's [3], Wu & Palmer's [4], Resnik's [6], Jiang & Conrath's [7] and Lin's [8] respectively.

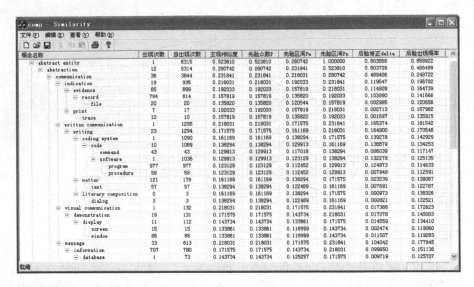

Fig. 2 The interface of experiment

Table 1 The Results of Similarity Computation

Concept 1	Concept 2	BES	HSO	LCH	WUP	RES	JCN	LIN
communication	indication	0.922	4	2.996	0.889	3.071	0.343	0.678
indication	evidence	0.948	4	2.996	0.909	5.985	3.526	0.977
program	procedure	0.956	5	2.590	0.900	11.77	0.108	0.423
screen	window	0.943	6	2.590	0.909	11.07	0.094	0.426
letter	character	0.978	4	2.996	0.941	8.721	3.677	0.985
memory	disk	0.956	5	2.590	0.900	7.916	0.455	0.878
symbol	code	0.712	3	2.079	0.667	3.072	0.107	0.397
character	program	0.642	3	2.079	0.778	5.129	0.188	0.659
key	cursor	0.927	4	2.303	0.842	4.368	0	0
message	signal	0.804	5	2.590	0.800	3.072	0.235	0.591

Table 2 The Result of Similarity Comparison

	Score	Rate
BES	42	93.3%
HSO	35	55.6%
LCH	32	71.1%
WUP	41	91.1%
RES	29	64.4%
JCN	35	77.8%
LIN	36	80.0%

In fact, it's difficult to compare the performance of similarity computation algorithms directly by their results [17], because all values are computed by different principles and it's hard to tell that "0.922" is a better similarity between "communication" and "indication" than "4". Here we propose a relative way: we choose two rows in Table 1, such as row "communication-indication" and "indicationevidence" and compare them. If more than half of these algorithms figure that the similarity between "communication-indication" is smaller than that of "indication-evidence", the majority win one score. Through this process we count the score of each algorithm, and regard them as the merit. The statistical results are shown in Table 2. From the experimental results, we can see our approach is the best.

5 Conclusions

Similarity is an important and fundamental concept in ontology mapping, services discovery and semantic retrieval and many other fields. This paper makes contributions in the following aspects. It presents a new similarity measure between concepts which integrate the subjective experience with the objective statistic. Subsequently a relative similar method to compare different similarity algorithms is also proposed. More importantly, it proposes an important idea and approach to achieve concept encountering probability based on Bayesian Estimation. Experiments prove the performance of our method is the best, although we also take some empirical assumptions as other applications, which cannot be proved reasonable.

References

[1] Klein, M., Bernstein, A.: Searching for Services on the Semantic Web Using Process Ontologies. In: The First Semantic Web Working Symposium, Stanford, CA, USA, pp. 159–172 (2001)
[2] Hirst, G., St-Onge, D.: Lexical chains as representations of context for the detection and correction of malapropisms. In: WordNet: An Electronic Lexical Database, pp. 305–332. MIT Press, Cambridge (1998)
[3] Leacock, C., Martin, C.: Combining local context with WordNet similarity for word sense identification. In: WordNet: An Electronic Lexical Database, pp. 265–283. MIT Press, Cambridge (1998)
[4] Wu, Z., Palmer, M.: Verb semantics and lexical selection. In: Proceedings of the 32nd Annual Meeting of the Associations for Computational Linguistics, Las Cruces, New Mexico, pp. 133–138 (1994)
[5] Tervsky, A.: Features of Similarity. Psychological Review 84(4), 327–352 (1977)
[6] Resnik, P.: Using information content to evaluate semantic similarity in a taxonomy. In: Proceedings of the 14th International Joint Conference on Artificial Intelligence, Montreal, Canada, pp. 448–453 (1995)
[7] Jiang, J.J., Conrath, D.W.: Semantic similarity based on corpus statistics and lexical taxonomy. In: Proceedings of International Conference on Research in Computational Linguistics, Taiwan, pp. 19–33 (1997)

 [8] Lin, D.: An information-theoretic definition of similarity. In: Proceedings of the 15th International Conference on Machine Learning, pp. 296–304. Morgan Kaufmann, San Francisco (1998)
 [9] Gruber, T.R.: A translation approach to portable ontology specification. Knowledge Acquisition 5(2), 1992–2201 (1993)
[10] W3C. OWL Web Ontology Language Reference, http://www.w3.org/TR/owl-ref
[11] Liu, Q., Li, s.: A Word Similarity Computing Method Based on HowNet. In: The 3rd Chinese Lexical Semantics Workshop, Taibei, pp. 59–76 (2002)
[12] Chen, X.: An Introduction to Mathematical Statics. Science Press, Beijing (1999)
[13] Fang, K., Xu, J.: Statistical Distribution. Science Press, Beijing (1987)
[14] Princeton University Cognitive Science Laboratory, WordNet – a lexical database for the English language, http://wordnet.princeton.edu
[15] Cormen, T.H., Leiserson, C.E., Rivest, R.L.: Introduction to Algorithms. MIT Press, Cambridge (2001)
[16] Pedersen, T.: Wordnet:Similarity, http://wnsimilarity.sourceforge.net
[17] Budanitsky, A., Hirst, G.: Semantic distance in WordNet: An experimental, application-oriented evaluation of five measures. In: Workshop on WordNet and Other Lexical Resources, Pittsburgh (2001)

What Make Democracy Possible: A Predictive Modeling Approach

Ghada Sharaf El Din, Carl Lee, and Moataz Fattah

Summary. Data Mining techniques have been successfully applied to many disciplines, especially in business intelligence, medical and health related problems. This paper presents a case study using data mining predictive modeling techniques for identifying important factors associated with the making of democracy. Democracy has been a paradoxical phenomenon; it has appeared in countries that defy the theoretically proposed prerequisites such as high level of economic development, low level of illiteracy, and strong middle class among others. While other countries that have these prerequisites barely witnessed successful democratic transitions. This discrepancy between theory and empirical observation calls for an investigation of factors beyond social & economic factors associated with democratization. A predictive modeling approach will be applied to develop models using political factors, social-economic factors, degree of development, etc. to predict an index derived from the institutional and normative aspects of democracy.

Keywords: Data Mining, Decision Tree, Democracy, Multilevel Logistic Regression, Neural Network.

1 Introduction

The applications of data mining techniques have been widely found in many disciplines, especially in business intelligence, medical and health, security and many others. This article presents a case study for identifying important factors associated with the making of democracy. Democracy has been a paradoxical phenomenon; it has appeared in countries that defy the theoretically proposed prerequisites such as high level of economic development, low level of illiteracy, and strong middle class among others. Countries like India, Bolivia, Senegal, Mali and Bangladesh, according to several modernization and democratization schools, should not have witnessed democratic transitions, yet they are currently democratic. While other countries (ex. Saudi Arabia, Belarus, Armenia, Egypt, and Iran) that have these prerequisites barely witnessed successful democratic transitions [3, 13]. This discrepancy between theory and empirical observation calls for a large scale study that would use panel data via allowing variation through cases and

Ghada Sharaf El Din, Carl Lee, and Moataz Fattah
Central Michigan University, Mt. Pleasant, Michigan USA
e-mail: Carl.lee@cmich.edu

R. Lee, N. Ishii (Eds.): Software Engineering, Artificial Intelligence, SCI 209, pp. 145–155.
springerlink.com   © Springer-Verlag Berlin Heidelberg 2009

time at the same time to discern the variables that are highly associated with the initiation and consolidation of democratic transition. In this project, we apply a predictive modeling approach to examine the political, social-economic factors, degree of development, etc. that can robustly predict the institutional and normative aspects of democracy.

2 Theoretical Framework of the Making of Democracy

A body of literature, modernization school, [1, 10, 11] that dominated political science for a couple of decades, proposed that societies advance as a whole economically, socially, technologically and politically. Any development in any field will have a spill-over effect on other areas. Yet, there are examples of countries that achieved high levels of economic growth (example Arab Gulf states), yet they still lag behind in political evolution. Others are relatively advanced politically while struggling in all other areas (India, Mali, Senegal and Bolivia are illustrative examples). These studies were often limited by the availability of data sources or the scope of the assumed framework. These discrepancies suggested to another school of political literature that democratization is about political actors' consequential choices rather than economic and social factors and prerequisites [9, 14]. Wejnert [14] studied the diffusion effect in the development of democracy and suggested that, when diffusion factors are included in the study, the contributions of social and economic factors become very small and the diffusion factors provide much more weights in explaining the development of democracy. In this study Wejnert [14] applied a hierarchical mixed model approach by taking into account both social, economic and diffusion factors into consideration. Her study suggests that further research is needed to combine as complete as possible the available data sources into a large scale study using the modern data mining techniques that are developed for handling high dimensional large data set. In this study, we expand our scope to include a more complete set of factors that are categorized as below: (1) external factors (2) political factors (3) social economic factors (4) technological development (5) Strength of private business sectors and (6) educational factors. Detailed discussion related to each category will be addressed in the data source section. The purpose of the study is to apply data mining predictive modeling techniques to investigate the important factors that are highly associated with the development of democracy using as complete as possible the available data sources focus on investigating the relative importance of all of these available factors including those were considered in the literature and many other which are yet to be studied.

3 The Data Sources and Data Preparation

The data source is obtained from the ICPSR [7]. As detailed in the description of the data source web site, the data consists of variables from six existing databases. The data is collected from 1800 to 2005. It has 312 variables and 15,655 observations for a total of 187 sovereign countries.

The response of interest is the degree of democracy index from 0 to 10 with 0 being no democracy and 10 being highest level of democracy. This index is proposed to consist of the institutional (party competition, free elections, peaceful alteration of power) and normative aspects of democracy (freedoms of belief, expression, association, and participation).The development and applications of this index are well documented in the literature e.g., Dahl [2 (p.38)], Jagger and Gurr [8], Wejnert [14]. Fig. 1 summarizes the average democracy index by year. The index shows clearly three waves of democratization. The first one brought democracy to Western Europe and Northern America in the 19th century. It was followed by a rise of dictatorships during the interwar period. The second wave began after World War II, but lost steam between 1962 and the mid-1970s. The latest wave began in 1974 and is still ongoing. Democratization of Latin America and post-Communist countries of Eastern Europe is part of this third wave [6]. The average yearly indices were lower than one during the early 1800's with the lowest average in 1817. In 1946 (after World II), the average was at 4.58. In 1994 (after the Cold War), the average was 5.33. The all time high occurred in 2001 with the average at 5.48. Fig. 2 summarizes the number of countries included in the data. For a given year, the minimum number of countries was 20 and the maximum number of countries was 158 in 1994. There was a drop occurred in 1996 from 158 to 132 and back to 153 in 1997.

Among the 312 variables, those variables not related to the purpose of this study are excluded immediately. A sequence of preliminary statistical assessments is conducted to screen out less relevant variables. The input variables selected for this study are those pass preliminary statistical assessment and the variables suggested by the theoretical framework. A total of 56 variables are chosen as our

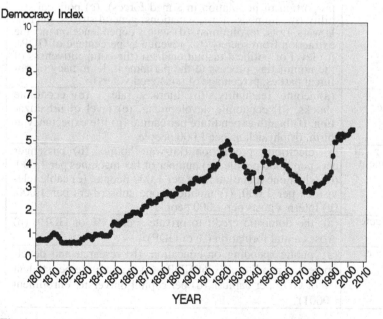

Fig. 1 Average Yearly Democracy Index from 1800 to 2005

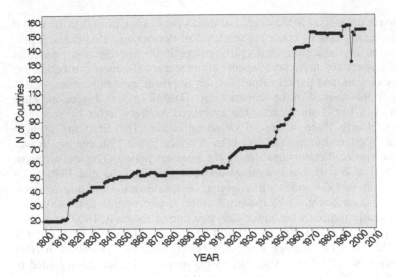

Fig. 2 Number of Countries in the Data set by Year from 1800 to 2005

Table 1 The Definition of Input Variables

External Factors	(a)engagement of countries in wars, (b) being occupied, (c) engagement in the global world (d) foreign direct investment net flows (% of GDP)
Political Factors	(a) power concentration (unitary vs. federal states), (b) militarization of society (percentage of defense expenditure and proportion of population in armed forces), (c) political stability (the number of assassinations, general strikes, guerrilla wars, riots, revolutions), (d) state's dependence on income extraction from society (tax revenue as percentage of GDP), (e) level of political institutionalism (the competitiveness of the nominating process to the parliament, legitimacy of political parties, percentage of registered voters).
Social Economic Status	(a)income inequality, (b) literacy rate, (c) economic shocks, (d) economic development, (e) level of urbanization, (f) health expenditure per capita, (g) life expectancy at birth, (h)physicians per 1,000 people
Technological Development	(a) electricity production (kil-watt hours), (b) passenger cars per 1,000 people, (c) number of fax machines per 1,000, (d) telephone mainline lines per 1,000 people, (e) cable televisions per 1,000, (f) mobile phone subscribers per 1,000, (g) Internet hosts per 1,000 people
Private Sector Status	(a) the domestic credit to private sector (% of GDP), (b) gross capital formation (% of GDP)
Education & Research	(a) public spending on education, (b) research and development expenditure (% of GDP), (c) expenditure per student (% of GDP per capita), (d) per capital university enrollment (.0001)

input variables (factors) in this study. Among the 56 variables, twenty eight of them have 50% or more missing data and are excluded from further analysis. The remaining variables are classified into six categories described in Table 1. Among these remaining variables, we perform missing data manipulation using the following imputation strategy: the missing cases of a variable are imputed so that the distribution of the imputed data is as close as possible to the un-imputed data of the variable. Furthermore, we conduct variable transformation using the following strategy:

- For interval-scale variables, the maximum normal transformation is employed so that the transformed variable is as close to normal as possible.
- For categorical variables, grouping rare levels is employed with the cutoff percentage being 0.5%.

SAS programs are written to create the subset of the data and to perform further data manipulation such as handling missing data.

4 Modeling Methodology

Three types of predictive modeling techniques are applied and compared to select the 'best' model using several model selection criteria. The modeling techniques used include (1) multilevel logistic regressions, (2) decision tree models, and (3) neural network models. The data are partitioned into Training (60%), Validation (30%) and Test (10%). The Training data is used to develop the models and Validation data is use to finalize the 'best' model for each modeling technique, and the Test data is held out to examine the performances among the best models of different modeling techniques for selecting the 'best' overall model. The SAS/ Enterprise Miner software is used for the data mining analysis. Fig. 3 illustrates the modeling methodology used in this study.

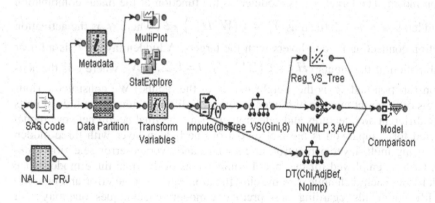

Fig. 3 The Modeling Strategy Used in the Study

A decision tree model is developed using both original and transformed variables. Maximizing the Chi-Square Statistic with Kass adjustment is applied for splitting the rules using the Training data set. The pruning is performed using the assessment criterion of minimizing the misclassification rate based on the validation data set. The missing data is treated as a separate category and is used in the modeling construction. The importance of input variables is determined by using the logworth measure.

Prior to conducting logistic regression and neural network models, the data is imputed and a variable selection process using decision tree based on Gini criterion is conducted. This serves as a preliminary screening of input variables. Several logistic regressions and neural network models are experimented. One final model from each modeling technique is selected in the final model development and comparison. The link function used for logistic regression is logit link function. The target variable is ordinal scale, and hence the multilevel logistic regression is constructed using the cumulative logits. The model is given as below, Let Y be the target, Democracy Index, with 11 levels (0 to 10) and X be the vector of input variables. The model is

$$P(Y = i \mid X) = \frac{\exp(\beta_{0i} + \beta' X)}{1 + \exp(\beta_{0i} + \beta' X)},$$

where i = 0,1,2 ...,10. The weights for the input variables are the same for each level of Y, and the bias terms β_{0i}'s are different. The cumulative logits model is also known as the proportional odds model, which is a common approach when the target is ordinal scale. A generalized logistic regression model by treating the democracy index target as a nominal scale is also conducted, and the results are found similar to the cumulative logits models. The cumulative logits logistic model is applied in order to maintain the numerically meaningful order of the democracy levels.

A neural network (NN) can be considered as a two-stage nonlinear or classification model. The target $Y_{i.}$ is modeled as the function of the linear combination of hidden layers $H_{i.}$ defined as $Y_{i.} = f\left(W'H_{i.}'\right) + \varepsilon$, where f is the activation function connecting hidden layers with the targets. A hidden layer H is a linear combination of the inputs: $H_{i.} = g\left(Z'X_{i.}'\right)$, $i = 1, 2, \ldots, N$, where g is the activation function and Z is the weight matrix of the inputs. We construct various models of different number of hidden layers with different number of hidden units using different architectures and selection criteria. The final neural network model selected for comparison is the NN model of one hidden layer with three hidden units using multi-layer perceptron architecture and average error selection criterion. Criteria employed for the model comparisons of the final three models (one model from each technique) are misclassification rate, average error and cumulative lift. For details regarding these predictive modeling techniques, one may refer to Sarma [12]. Han and Kamber [5], Giudici [4], and others.

5 Results and Discussions

The selected models from the three modeling techniques are compared based on three criteria: (1) the misclassification rate (MS Rate), (2) the root average square error and (3) the cumulative lift at the top 10% level. The results are given in Table 2. It indicates the decision tree model is the best using all three criteria. The misclassification rate is at 18% for decision tree model, while the other two models are over 33%.

Table 2 The Model Comparison

Model setting	MS Rate	Root ASE	Cum. Lift (10%)
Decision Tree	0.180	0.149	6.40
Logistic Reg.	0.332	0.194	6.14
Neural Network	0.369	0.204	6.28

We select the decision tree model as our final 'best' model. The remaining results and discussions are based on the finding from the decision tree model. The important input variables selected by the decision tree model are given the Table 3.

Table 3 The Selected Important Input from the Decision Tree Model

Input Variables	Relative Importance
Regulation of participation (PARREG)	1.00
Executive Recruitment Competition(XRCOMP)	.952
% Urban population (100,000), shifted (URBAN1CA)	.398
%labor force in industry & service interpolated (EXP_LABOIN_1)	.393
Parliamentary responsibility (PARLRES)	.346
Geographic region (REGION)	.247
Centralization of state authority (CENTRAL)	.194
Proportion of population in armed forces (.0001) (MILITPOP)	.163
% Defense expenditure (DEFENEXP)	.138
Gross capital formation (% of GDP) (INVESTDM)	.131
Executive Recruitment Openness (XROPEN)	.114
Per capita university enrollment (.0001) (PWR_UNIVCA)	.102
% Registered voters (PWR_VOTERPOP)	.101
Number of anti-government peaceful demonstrations (DEMONST)	.036

Geographical regions have different levels of democracy development. Most stable democracies are concentrated in certain regions mainly North America and Western Europe with clear democratic deficit in the Middle East, Africa, West and Central Asia. Political scientists and students of democratization have always been debating the priority of socio-economic factors over institution-crafting factors. In this research we included all the possible factors, socio-economic, cultural, institutional, and others to let the decision tree and empirical data discern which are more capable of helping us in understanding what make some polities more democratic than others.

5.1 General Finding

As a general finding, the preponderance of evidence suggests that the institutional factors are much more important in explaining our dependent variable, the democracy index. Six of the 14 factors are institution-related including the top two factors in the decision tree and only four of the 14 factors are socio-economic.

5.2 The Political Relevance of Institutional Factors

Institutional factors indicate that they pertain to the level of efficiency in institution crafting and management of conflicting demands. For example, countries may end up in civil wars because of poor and ambiguous relationships between the parliament, the executive, bureaucracy, judiciary branches. Examples range from Somalia, Yugoslavia, Lebanon among many others. According to this model the *regulation of political participation* (PARREG) operationalized as a function of the degree to which there are binding rules on when, whether, and how political preferences are expressed, is the most important factor in predicting institutional democracy. The essence of democracy is to set up the appropriate rules to manage and regulate participation and competition of citizens and groups over who gets what, when and how. This indicates that it is mandatory for democracies to take roots and to become consolidated to be coupled with an efficient institution-crafting that will ensure that relatively stable and enduring groups regularly compete for political influence with little use of coercion).

The second variable that is suggested by the model is *the competitiveness of executive recruitment* (XRCOMP). The competitiveness of executive recruitment means that democracy takes roots in polities that give adequate attention to the process of choosing their executives through competition among two or more parties or political coalition rather than inheritance or having one-party system. Polities that fail in coming up with a system that would allow fair completion and participation of its citizens and political forces, run the risk of being non-democratic.

The third relevant variable in explaining why certain polities are democratic while others are not is Parliamentary responsibility (PARLRES). Democracies enjoy a high level of "checks and balances" that ensure that the parliament oversees and monitors the wrongdoing of the executive branch and bureaucracy. Without this institutional element, the possibility of democratic stability is slim. The final three institutional variables that increase the chance of having stable democracies are related to: the nature of the *relationship between the central government and its peripheries* with federalism increasing the possibility of democracy than unitary regimes (CENTRAL)), the *executive recruitment openness* with more support for democracy in regimes that allow more open competition for filling executive offices rather than designation or favoritism (XROPEN), and the *percentage of registered voters* (PWR_VOTERPOP) indicating that the more the regimes is capable of convincing its citizens to be politically active, the more the chances of establishing a stable democracy.

5.3 Socio-economic Factors

Four socio-economic factors appear to be the relevant in explaining why some polities are more democratic than others: (1) *The level of urbanization* (the more people live in cities compared to rural areas, the higher the chance of democratic stability (URBAN1CA). (2) *%labor force in industry & service interpolated* (EXP_LABOIN_1). The higher the number of people who work in industrial and services sectors compared to agriculture and hunting, the more the possibility of democratization. (3) *Gross capital formation* (% of GDP) (INVESTDM). The richer the country is (measured by its gross capital formation), the more it tends to become democratic given the desire of its citizens to convert their educational and intellectual skills into political rights. (4) *Per capita university enrollment (.0001)* (PWR_UNIVCA). The more there are youths in universities or with college degrees, the more the chances of democratization. Taken together, these four variables indicate that the stronger the middle class in a country, the more likely it is to democratize and to be democratically stable which supports a political argument that: "no middle class, no democracy".

5.4 Arm Force Related Variables

Two variables that are not necessarily related to democratization yet very relevant for the security and stability of any polity appear in the model: *Proportion of population in armed forces and the percentage of defense expenditure.* Historically, strong democratic countries tend to have high proportions of armed forces and defense expenditure in order to maintain security and stability. The same is true that a strong dictatorship or unitary party controlled countries also have high percentage of population in armed forces and high percentage of defense expenditure in order to maintain the dictatorship or unitary party control.

5.5 Anti-governmental Demonstrations

It is not surprising that anti-governmental demonstration is an important factor impacting the democracy development. The progression from dictatorship to democracy often entails more freedom of expression, association and demonstration In democratic countries, such demonstrations are regulated by the law. In dictatorship or unitary party controlled government, such movements are most likely not allowed. The reaction from government is often by using force to crack down the incidences. This becomes a cycle of anti-government demonstration by citizens and cracking the demonstration by police forces. Such cycles often lead to more restricted ruling or moving towards democracy.

6 Conclusion

In conclusion, this article presents a case study of applying three different predictive modeling techniques that are common in data mining, decision tree, logistic regression and neural network, to investigate the insights of factors associated with the making of democracy using over 200 years of data. The decision tree model is found the 'best' model using three criteria: misclassification rate, cumulative lift and root average squared error. The misclassification rate based on the best model is at 18%. Among the 56 considered input variables, fourteen variables are found important. Findings in the literature have focused more on the social economic factors [1, 10, 11]. Recent studies by Wejnert [14] suggested that the diffusion factors are more important than social-economic factors using hierarchical mixed modeling approach. Our study takes one step further by taking into account all available input variables and years of data. Our findings suggest that there are four categories of factors that have important contribution to the making of democracy. The most important category is the political relevance of institutional factors followed by social economic factors, and military related factors as well as the anti-governmental demonstration activities. The weakness of this study is that the study does not take the time series into consideration. Further research by considering the impact of time lag of input variables to the development of democracy is needed. Another interesting research area is to predict the amount of time takes to develop democracy given a certain starting conditions of the input variables.

References

1. Bollen, K.: World System Position, Dependency, and Democracy: The Cross-National Evidence. American Sociological Review 48, 486–497 (1983)
2. Dahl, R.: On Democracy. Yale University Press (1998)
3. Fattah, M.: Democratic Values in the Muslim World. Lynn Rienner Publications (2006)
4. Giudici, P.: Applied Data Mining. John Wiley & Sons, Inc., Chichester (2003)
5. Han, J., Kamber, M.: Data Mining Concepts and Techniques. Morgan Kaufman Publishers, San Francisco (2001)
6. Huntington, S.P.: Democratization in the Late Twentieth Century. University of Oklahoma Press, Norman (1991)
7. Inter-University Consortium for Political and Social Research, #20440, http://www.icpsr.umich.edu/~cocoon/ICPSR/STUDY/20440.xml
8. Jagger, K., Gurr, T.R.: Tracking Democracy's Third Wave with the Policy III Data. Journal of Peace Research 32, 469–482 (1995)
9. Linz, J.J., Alfred, S.: Problems of Democratic Transition and Consolidation: Southern Europe, South America and Post-Communist Europe, p. 113. The Johns Hopkins University Press (1996)

10. Przeworski, A., Alvarez, M., Cheibub, J.A., Limongi, F.: Democracy and Development. Cambridge University Press, Cambridge (2000)
11. Rustow, D.A.: Transitions to Democracy: Toward a Dynamic Model. Comparative Politics 2, 337–364 (1970)
12. Sarma, K.: Predictive Modeling With SAS Enterprise Miner: Practical Solutions for Business Applications. SAS Publishers (2007)
13. Tessler, M.: Do Islamic Orientations Influence Attitudes Toward Democracy in the Arab World? Evidence from Egypt, Jordan, Morocco, and Algeria. International Journal of Comparative Sociology 43(3-5), 229–249 (2002)
14. Wejnert, B.: Diffusion, development and democracy, 1800-1999. American Sociological Review 70(1), 53–81 (2005)

Blog Summarization for Blog Mining

Mohsen Jafari Asbagh, Mohsen Sayyadi, and Hassan Abolhassani

Abstract. Although dimension reduction techniques for text documents can be used for preprocessing of blogs, these techniques will be more effective if they deal with the nature of the blogs properly. In this paper we propose a shallow summarization method for blogs as a preprocessing step for blog mining which benefits from specific characteristics of the blogs including blog themes, time interval between posts, and body-title composition of the posts. We use our method for summarizing a collection of Persian blogs from PersianBlog hosting and investigate its influence on blog clustering.

Keywords: Summarization, Blog Summarization, Blog Mining, Dimension Reduction, Blog Preprocessing.

1 Introduction

Blog space on the Web, like other Web pages, can be considered as a huge source of information. Number of blogs has been raised tremendously in recent years and still keeps growing. We cannot call a blog as just a Web page because each blog usually consists of a large number of posts which cover various topics. Therefore, a blog can be seen as a page of pages. In order to extract information from such a big data collection, automatic analysis and discovery process is needed. Data mining techniques can be applied to blogs in order to discover interesting knowledge from them which we refer by blog mining during the remainder of this paper. Taking into account the big count of blogs, large number of posts in each blog, and diversity in the topics of posts, it is obvious that mining in this huge data storage is a challenging work. Therefore, an appropriate preprocessing can increase the quality of gained knowledge as well as decrease the complexity of the mining process.

One of the effective preprocessing tasks in mining process is to reduce the dimensionality of data collection. There are numerous dimension reduction

Mohsen Jafari Asbagh and Mohsen Sayyadi
406, Khodro Building, Sharif University, Azadi Ave., Tehran, Iran
e-mail: {m_jafari, m_sayyadi}@ce.sharif.edu

Hassan Abolhassani
222, Computer Engineering Department, Sharif University, Azadi Ave., Tehran, Iran
e-mail: abolhassani@sharif.ir

R. Lee, N. Ishii (Eds.): Software Engineering, Artificial Intelligence, SCI 209, pp. 157–167.
springerlink.com   © Springer-Verlag Berlin Heidelberg 2009

techniques for text documents that can be used to reduce the size of the vector of a text document in vector space model [1, 2]. These techniques model a document in a low-dimensional space using statistical and analytical methods. The drawback of these techniques is that they do not pay attention to the position of words in a document; relation of a word to main topic of the document; and the various themes in a document which is a usual phenomenon in Web pages and blogs especially. Text summarization techniques that use *shallow* model have potential to regard all the aforementioned factors. Shallow approach is one of the two main approaches of text summarization according to the level in *Linguistic Space* which usually does not venture beyond a syntactic level of representation [3]. *Deeper* approach is the other one that usually assumes at least a sentential semantics level of representation [3]. A survey of text summarization methods can be found in [4].

In this paper we propose a shallow summarization method for blogs as a preprocessing step for blog mining. Although the output of summarization techniques is mainly supposed to be read by humans, but, since they reduce the size of a text document considerably, they can serve as preprocessing methods for mining text documents. Summarization algorithms usually focus on sentences in documents and identify the important sentences to represent the document, but our method uses posts and cluster of posts to score the important words. Also, identifying representative sentences is not its concern and it just selects representative terms and weighs them based on importance of blog themes.

When developing a blog-specific summarization method, one has to appropriately handle specific characteristics of the blogs including blog themes, time interval between posts, body-title composition of the posts, link structure, and comments. In our method we take benefits of first three items and we are going to include two last items in future works.

The following two main reasons motivated us for developing such a method:

- Dimension reduction techniques and Summarization methods that are applied to blogs should be developed according to special characteristics and dynamics of blogs.
- Persian text processing generally suffers from low quality. One of the main reasons is the lack of a comprehensive solution for stemming of verbs and nouns. Hence, vector space in Persian text mining is too large and consequently the process is very time consuming and leads to weak results. We believe that our method in the absence of stemming can enormously decrease the size of the vector space and affect the Persian text mining process positively.

The rest of this paper is organized as follows. Section 2 represents some related works. In section 3 we explain steps of our algorithm in detail. Section 4 includes details of our experimental evaluation which briefly describes how data collection was prepared, how experiment was leaded, and what results were achieved. Finally, in section 5 we conclude our work and state the future directions in our research.

2 Related Works

Among numerous shallow summarization methods in the literature, Maximum Marginal Relevance Multi Document (MMR-MD) [5] and MEAD [6] are two methods that our idea is most similar to. They analyze clusters of a document to summarize it. MMR-MD is a purely extractive summarization method that is based on Maximal Marginal Relevance concept proposed for information retrieval. It can accommodate a number of criteria for sentence selection such as content words, chronological order, query/topic similarity, anti-redundancy and pronoun penalty. MEAD is a sentence level extractive summarizer that takes document clusters as input. Documents are represented using term frequency-inverse document frequency (TF-IDF) of scores of words. IDF value is computed based on the entire corpus. The summarizer takes already clustered documents as input. Each cluster is considered as a theme. The theme is represented by words with top ranking TF-IDF scores in that cluster. Sentence selection is based on similarity of the sentences to the theme of the cluster.

Recently much work has been done on Web page summarization. OCELOT [7] is a prototype system for automatically summarizing Web pages. It uses probabilistic models to generate the "gist" of a Web page. The models used are automatically obtained from a collection of human-summarized Web pages. Sun et al used LSA and Luhn's sentence selection methods to summarize a Web page using clickthrough data [8]. In their work, clickthrough data is supposed to provide some human understanding about Web pages. Also some research has been done to enhance the level of accuracy of Web page classification based on Web summarization [9].

On the other side, Blogs have received much attention from researchers in recent years. There already exist several techniques for spam blog post detection, blog posts tagging and opinion mining. Hu and Liu proposed an opinion summarization of products, categorized by the opinion polarity [10]. They then illustrated an opinion summarization of bar graph style, categorized by product features [11].

Nevertheless, very few studies on blog post summarization have been reported. Zhou et al viewed a blog post as a summary of online news articles it linked to, with added personal opinions [12]. A summary is generated by deleting sentences from the blog post that are not relevant to its linked news articles. Comments associated with blog posts were however not used. There are also some existing researches based on comments. In [13] a new solution is proposed which first derives representative words from comments and then selects sentences which include those words. They believe that reading comments have serious affects on one's understanding about a blog post.

3 Summarization Algorithm

In this section we are going to describe our algorithm for blog summarization. Our method mainly consists of three steps. First, we cluster posts of each blog in order to break it down into groups of posts which we guess each group has posts with the same topic. According to [14], we expect that these groups have the same themes. In their paper, authors by theme refer to common things that might not be explicitly stated in the content but two posts having the same underlying theme are highly related. We name this step as *Theme Identification*. After this step, in *Score Computation* step we compute a score for each group based on three quality measures which each one indicates a factor that influences the excellence of a group. *Word Selection* is the last step of the algorithm. In this step we select a bunch of words among these groups as the representative of corresponding blog in blog mining task. In the remaining of this section we explain these three steps in more details.

3.1 Theme Identification

The first step toward theme identification is grouping of blog posts. Any clustering algorithm can be used for grouping but, since the count of posts in a blog is not usually a large number the simplest way is to construct the mutual similarity matrix of posts. We define δ_{sim}, a number between 0 and 1, as a threshold which two posts with similarity greater than it will be put in the same group.

The similarity of two posts arises from the similarity between their titles as well as bodies. Usually posts in a blog don't have the title. Using a query in Persian-Blog database, we observed that almost 60 percent of posts lack the title. In the other hands, there are plenty of posts with the same or highly similar titles that their bodies have nothing in common. Since our aim of summarization is a pre-processing step for blog mining, we expect that finally remained words be good representatives for whole blog. Therefore, although title similarity is important in many applications, we give it credit only when the bodies also have a good similarity. In other words, similarity of body controls the impact of the similarity of the titles on overall similarity between two posts. The similarity measure of two posts, $S_{p1,p2}$, is stated in equation (1). In this formula $SB_{p1,p2}$ and $ST_{p1,p2}$ represent body similarity and title similarity of two posts, respectively. Multiplication of $SB_{p1,p2}$ in $ST_{p1,p2}$ guarantees that high similarity in title can be of importance if the body similarity is high as well. Coefficients u_1 and u_2 are two real numbers that determine importance of two factors in summation.

$$S_{p1,p2} = u_1 * SB_{p1,p2} + u_2 * SB_{p1,p2} * ST_{p1,p2} \ , u_1, u_2 \in [0,1] \wedge u_1 + u_2 = 1 \qquad (1)$$

Even though blog posts are actually a part of a Web page and therefore are stored in HTML format, for simplicity we ignore HTML tags and just use plain text for computing $SB_{p1,p2}$ and $ST_{p1,p2}$.

After finding the clusters of posts, clusters with small population are being removed. The reason behind this removal is the fact that themes with very small number of posts can not be a part of main interests of the owner of the blog. For this purpose, we define another threshold, δ_{pop}, as an integer number that the posts of the clusters with populations less than it will be removed from blog posts.

3.2 Score Computation

After the clustering of posts and removing small clusters, for each cluster we compute a score which indicates the excellence of its underlying theme. The higher the score of a cluster, the better representative its posts are for whole blog. We define cluster score based on three quantitative factors of average time interval between posts, cluster population, and average post similarity ($S_{p1,p2}$) as defined in equation (1). Equation (2) shows the definition of the score of cluster c which we name it SCR_c.

$$SCR_c = w_1 * (1 - T_c) + w_2 * C_c + w_3 * S_c \quad , w_1, w_2, w_3 \in [0,1] \wedge w_1 + w_2 + w_3 = 1 \qquad (2)$$

T_c represents the average time interval between posts of a cluster. This value is normalized to the interval of [0,1] for all clusters of a blog. The small value for this factor shows that blog owner writes posts with corresponding theme frequently and this shows that this theme has the potential to be one of the interests of the author.

C_c is the normalized -to interval [0,1]- count of the members of cluster c. We guess that clusters with bigger populations be of more importance to blog author.

S_c is the average mutual post similarity (equation (1)) of cluster members that like other two factors is normalized to interval [0,1] for all clusters. The bigger the value of S_c for a cluster, the stronger its underlying theme is and consequently the better representative it is for the blog.

w_1, w_2, and w_3 are used for weight adjustment between above three parameters. The summing up to 1 for these three parameters results in a value between 0 and 1 for SCR_c.

3.3 Word Selection

The last step of our algorithm involves selection of appropriate words as blog representatives based on computed scores for each cluster in previous step. Since our method summarizes blogs in order to be used for mining and the result will not be read by human, we just select words and do not try to identify important sentences. In order to select representative words, in first step, term frequency vector is computed for each cluster. We define another parameter which is a frequency threshold for identifying frequent terms in a cluster. We refer to this parameter by δ_{tf}. Using this parameter, frequent terms are those terms that their occurrence frequency in a cluster exceeds δ_{tf}. Non-frequent terms will be discarded and no score will be computed for them.

After determination of frequent terms for each cluster, the score of each term in the blog is computed by multiplying the frequency count of that term by the score of its containing cluster. If a term is frequent in more than one cluster, the biggest score will be selected for it. The formula for computing the score of a term which at least is frequent in one cluster is shown in equation (3). We refer to this score by WGH_t for term t. $TF_{t,c}$ indicates the term frequency of term t in cluster c.

The final result of this step is a list of unique words along with their scores for each blog which can be used either as a weighted term vector in vector model or for selecting top scored words in the case that there is a limitation on the vector size and the number of words which we are allowed to be selected for each blog is less than the number of scored terms.

$$WGH_t = Max\left(TF_{t,c} * SCR_c\right), \forall c : TF_{t,c} \geq \delta_{tf} \qquad (3)$$

4 Experimental Evaluation

In order to evaluate our algorithm, we decided to do some experiments on existing blogs on PersianBlog hosting. We investigated effectiveness of our algorithm on the clustering task and achieved results seem promising. For evaluating the clustering solution we chose F-measure and purity evaluation measures. Since there is no established method of stemming for Persian texts, term vectors are usually higher in dimension and sparser than those of English texts with almost same text size. This causes that internal criteria for clustering evaluation such as scattering criterion lack a good precision and therefore we attempted to use external criteria. A survey of clustering evaluation methods can be found in [15] and [16]. In next subsections we describe evaluation process in more details.

4.1 Blog Tagging

In order to compute F-measure and purity for clustering, we need to have the label of each blog. In PersianBlog collection there is no label associated with blogs, so, we had to give a class label for involved blogs in experiment by means of human judgment. We randomly chose 500 blogs including 160 old blogs (older than 4 years) and 240 newer ones. Selected blogs were tagged twice independently by two different persons and then the tags integrated for each blog. Two points must be noted regarding our tagging task:

- There was no predetermined tag set. The tag set grew up in the process of tagging.
- Since blogs have posts with different topics, it was really hard or impossible to specify just one tag for some blogs. Thus we decided to use three tags instead of one to overcome the problem. Hence, blogs got one, two, or three tags depending of their level of specificity.

At the end of tagging task, our tag set had 64 items which was pretty big for a data collection with 500 members. Most of these tags were just used once or twice and they couldn't be treated as a category. Therefore, we refined results by doing some generalizations between tags as well as removing rarely used tags and corresponding blogs. The final result was a collection of 394 blogs tagged using a tag set of 11 members. Table 1 represents these tags along with the number of blogs getting tagged using them.

Table 1 Grouping of blogs based on their labels

Label	Count	Label	Count
Spam	227	Information Technology	22
Love	102	Humanity	19
Politics	59	Football	16
Religion	38	Science	12
Psychology	28	Nuclear Power	6
Theology	27		

4.2 Leading the Experiment

We started our experiment with some preprocessing tasks. Unfortunately, there is no effective stemming tool for Persian, so we skipped stemming part in our experiment. However, the proposed method can be used on any language and when using for a language for which there is a stemming tool, stemming step can be included in the process. We took two steps: first, because we had no concern about HTML tags, we removed them from posts. Then, we removed Persian stop-words published in [17].

In order to evaluate our method we clustered two collections of blogs. First collection which we call original collection included 394 tagged blogs which HTML tags and Persian stop-words had been removed from their posts. The other collection was the result of applying our method on the original collection which we refer to by summarized collection. We used Cluto [18] clustering tool to cluster these collections. This tool requires that term vectors of documents to be clustered be given in a matrix, one row per document. Therefore, for the original collection we first concatenated all posts of each blog and then formed the vector matrix of the whole collection. For the matrix of this collection we used the frequency of occurrence of each term in a blog as its weight in the term vector of that blog. For constructing the matrix of summarized collection we used the score of each term computed by equation (3) for each blog.

Furthermore, we compared performance of our method to LSI [1] dimension reduction technique for text documents. We applied LSI on original collection and chose 50 first dimensions for clustering purpose.

4.3 Evaluation Process and Results

As mentioned earlier, because of the diversity of the topics in posts of a blog, we decided to assign two or three labels for some blogs which assigning only one label for them was erroneous. Therefore, when trying to compute F-measure and purity of clustering we should appropriately deal with multi-label blogs. As the solution, we pick a label of two or three labels in favor of more frequent tag in corresponding cluster. To be more specific, we choose the label which the count of blogs with that tag in the owner cluster is bigger than count of those with other tags. Algorithm TagAssignment in Fig. 1 shows the pseudo code for the steps of this procedure.

```
ALGORITHM TagAssignment
    CM: A matrix representing the number of documents of each label in each cluster
Begin
    For each cluster Cᵢ do
        For each single-label blog B in cluster Cᵢ with label Lⱼ do
            CM<Cᵢ,Lⱼ> := CM<Cᵢ,Lⱼ> + 1
            Mark B as read
        While there exists unread blog in cluster Cᵢ do
            Create LC as an array of length label_count and initialize each item to 0
            For each unread blog B in cluster Cᵢ do
                For each label Lⱼ of blog B do
                    LC<Lⱼ> = LC<Lⱼ> + 1
                Pick a label Lᴄ which yields the maximum value for CM<Cᵢ,Lᴄ> + LC<Lᴄ>
                CM<Cᵢ,Lᴄ> := CM<Cᵢ,Lᴄ> + LC<Lᴄ>
                Mark each unread blog in Cᵢ which has Lᴄ as one of its labels as read
End
```

Fig. 1 TagAssaignment Algorithm

Our algorithm takes the following actions for each cluster which we refer to by target cluster. First, it counts the labels of those blogs with one label and then for the rest which obviously have more than one label first keeps count of occurrences of each label for target cluster in a separate array and then chooses a label that in the case of addition to main matrix of counts yields the largest number. After choosing the appropriate label, it removes those blogs having that label by marking them as read and iterates the label choosing steps until no more unread blog remains in target cluster.

We tested our algorithm with various values for parameters. Here, we present the average case results in Fig. 2 which are yielded using the values indicated in Table 2.

Fig. 2-a and Fig. 2-b respectively compare F-measure and purity of clustering of summarized and original collection. It can be seen that the quality of clustering of original collection, especially in terms of F-measure, is very low. We achieve in average 42% improvement in F-measure and 12% promotion in purity by using our summarization method which seems very promising in particular for Persian language which lack of stemming results in a low quality in almost every task of information retrieval and mining on Persian documents.

Table 2 List of parameters and their used values

Parameter	Value	Parameter	Value
δ_{sim}	0.15	w_1	0.15
u_1	0.5	w_2	0.35
u_2	0.5	w_3	0.5
δ_{pop}	3	δ_{tf}	2

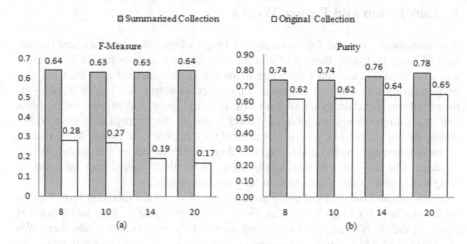

Fig. 2 a) F-measure b) Purity, of clustering of summarized and original collections for different number of clusters

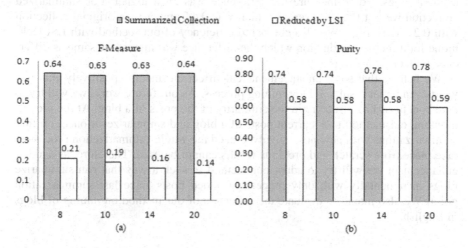

Fig. 3 Comparison of summarization method with LSI in terms of a) F-measure b) Purity

Fig. 3 shows the comparison of our summarization method to LSI dimension reduction technique. Results look same as the case that LSI has not been used. This huge difference in results arises from the fact that our method acts with regard to specific characteristics of blogs and it can be stated that when trying to apply a dimension reduction techniques to documents with specific attributes like blogs, it is better to use a method that deals properly with those attributes instead of using general dimension reduction techniques which solely use some statistical and mathematical analysis to achieve the result.

5 Conclusion and Future Works

We developed a method for summarizing blogs which takes into account the nature of blogs including themes of a blog, body-subject composition of blog posts, and time interval between posts. The main aim of this method is to be used as a preprocessing step for blog mining and therefore its output is a bag of words. Our proposed method determines themes of a blog by doing a clustering on its posts and then computes a score for each cluster based on the average time interval between its posts, its size, and average similarity between its posts. After that, it identifies frequent words and computes their scores based on the score of their clusters. The list of frequent words along with their scores can be used as weighted vector of the whole blog in blog mining process.

We evaluated our method using some of the blogs of PersianBlog blog hosting which is a host for blogs in Persian. First, we assigned one to three labels for 394 blogs by carefully reading its posts and then did our experiments over them. We clustered those blogs using summarized version as well as original ones and measured F-measure and purity for both clustering results. Clustering of summarized blogs had 76% purity in average which was 13% higher than clustering with original blogs. For F-measure the difference was even higher. For summarized collection we got 0.64 in average which was 0.44 higher than original collection with 0.22 in average. We also compared efficiency of our method with LSI technique for dimension reduction which the difference was almost the same as difference with original blogs.

We believe that some enhancements can affect our method positively and may yield even better results in blog mining process. As our future work, we will try to take benefit of link structure in the similarity of the posts of a blog. At the present moment, our method uses current posts of a blog and summarizes it once. Further summarizations with new posts involved need the whole summarization process to take place from scratch and previous results cannot be used. Therefore, as another enhancement we will try to adapt our method in such a way that can summarize blogs incrementally with only processing added posts since last summarization point. We also intend to evaluate the influence of our method on mining of blogs in English.

References

[1] Tang, B., Shepherd, M., Milios, E., Heywood, M.: Comparing and combining dimension reduction techniques for efficient text clustering. In: Proceeding of SIAM International Workshop on Feature Selection for Data Mining, pp. 17–26 (2005)
[2] Molina, L.C., Belanche, L., Nebot, A.: Feature selection algorithms: a survey and experimental evaluation. In: Proceeding of ICDM 2002, pp. 306–313 (2002)
[3] Mani, I.: Automatic summarization. John Benjamins Publishing, Amsterdam (2001)
[4] Jones, K.S.: Automatic summarizing: factors and directions. In: Advances in automatic text summarization. MIT Press, Cambridge (1999)

[5] Carbonell, J., Goldstein, J.: The use of MMR, diversity-based reranking for reordering documents and producing summaries. In: Proceeding of ACM SIGIR 1998, pp. 335–336 (1998)

[6] Radev, D., Allison, T., Blair-Goldensohn, S., Blitzer, J.C., Elebi, A., Dimitrov, S., Drabek, E., Hakim, A., Lam, W., Liu, D., Otterbacher, J., Qi, H., Saggion, H., Teufel, S., Topper, M., Winkel, A., Zhang, Z.: MEAD - a platform for multidocument multilingual text summarization. In: Proceeding of LREC 2004 (2004)

[7] Berger, A.L., Mittal, V.O.: OCELOT: a system for summarizing Web pages. In: Proceeding of 23rd Annual International ACM SIGIR Conference on Research and Development in Information Retrieval, pp. 144–151 (2000)

[8] Sun, J.T., Shen, D., Zeng, H.J., Yang, Q., Lu, Y., Chen, Z.: Web-page summarization using clickthrough data. In: Proceeding of SIGIR 2005, pp. 194–201 (2005)

[9] Shen, D., Chen, Z., Yang, Q., Zeng, H.J., Zhang, B., Lu, Y., Ma, W.Y.: Web-page classification through summarization. In: Proceeding of 27th Annual International ACM SIGIR Conference on Research and Development in Information Retrieval, pp. 242–249 (2004)

[10] Minqing, H., Bing, L.: Mining and summarizing customer reviews. In: Proceeding of SIGKDD 2004, pp. 168–177 (2004)

[11] Ku, L.W., Liang, Y.T., Chen, H.H.: Opinion extraction, summarization and tracking in news and blog corpora. In: Proceeding of AAAI-CAAW 2006 (2006)

[12] Zhou, L., Hovy, E.: On the summarization of dynamically introduced information: online discussions and blogs. In: Proceeding of AAAI-CAAW 2006 (2006)

[13] Hu, M., Sun, A., Lim, E.P.: Comments-oriented blog summarization by sentence extraction. In: Proceeding of CIKM 2007, pp. 901–904 (2007)

[14] Lin, Y.R., Sundaram, H.: Blog antenna: summarization of personal blog temporal dynamics based on self-similarity factorization. In: Proceeding of International Conference on Multimedia and Expo. (ICME 2007), pp. 540–543 (2007)

[15] Amigo, E., Gonzalo, J., Artiles, J., Verdejo, F.: A comparison of extrinsic clustering evaluation metrics based on formal constraints. In: Information Retrieval. Springer, Heidelberg (2008)

[16] He, J., Tan, A.H., Tan, C.L., Sung, S.Y.: On quantitative evaluation of clustering systems. In: Wu, W., Xiong, H., Shekhar, S. (eds.) Clustering and Information Retrieval, pp. 105–133. Kluwer Academic Publishers, Dordrecht (2004)

[17] Taghva, K., Beckley, R., Sadeh, M.: A list of farsi stopwords. Technical Report 2003-01, Information Science Research Institute, University of Nevada, Las Vegas (2003)

[18] Karypis, G.: CLUTO: a clustering toolkit. Technical Report 02-017, University of Minnesota (2002)

An Approach Using Formal Concept Analysis to Object Extraction in Legacy Code

Chia-Chu Chiang and Roger Lee

Summary. Formal concept analysis is a well-known technique for identifying groups of objects in software with common sets of attributes. The objects can be procedures, functions, and modules. In this paper, we are proposing an approach using formal concept analysis to extract objects in a non-object oriented programs. We define an object as a set of data and its associated methods encapsulated in a class. Thus, we start with a data structure and apply a program slicing technique to extract the code from the legacy code corresponding to the data. The data and statements are then collected into an object. This research is a collaborative work in which we are trying to draw the benefit by using formal concept analysis and apply the analysis results to the field of software maintenance.

1 Introduction

Many existing applications are written in non-object oriented programming languages such as C, PL/1, Ada, COBOL, and Pascal. We all remember how companies were struggling for pursuing solutions to ensure that their applications were Y2K compliant. Legacy applications are usually difficult to maintain due to software evolution [3]. It is very common to see legacy code contain dead code, clones, and goto statements that deteriorate the quality of the code. Several commercial tools have been developed to help remove dead code and clones from legacy programs. Several restructuring tools can be used to remove the goto statements in the programs also. These tools definitely improve the quality of the programs. However, there are still rooms for us to enhance the quality of legacy systems.

This paper deals with a problem of low cohesion in legacy applications. Low cohesive code indicates design problems such as lack of encapsulation or abstraction. The goal of this research is to idenity objects in a procedural code and extract the objects for software reuse. The results of re-enginnering procedural code into objects provide encapsulation and abstraction to the systems.

Chia-Chu Chiang
Department of Computer Science
University of Arkansas at Little Rock, 2801 South University, Little Rock, Arkansas 72204-1099, USA
e-mail: cxchiang@ualr.edu

Roger Lee
Department of Computer Science
Central Michigan University, Mount Pleasant, MI 48859, USA
e-mail: lee@cps.cmich.edu

R. Lee, N. Ishii (Eds.): Software Engineering, Artificial Intelligence, SCI 209, pp. 169–178.
springerlink.com   © Springer-Verlag Berlin Heidelberg 2009

The paper is organized as follows. Related work is in Section 2. The formal concept analysis is introduced in Section 3. The mathematical foundation of formal concept analysis is also described in this section with the illustration of examples. The approch of engineering procedural code into objects is briefly overviewed in Section 4. The details of each step of the approach are explained in Sections 5, 6, and 7. Several lessons from our experiences in developing this approach are presented in Section 8. Finally the paper is summarized in Section 9.

2 Related Work

Concept analysis has been applied to many kinds of problems. Work on the application of concept analysis in the area of software engineering has been studied since 1993. Areas include the use of concept analysis for finding inferences between configurations [11, 16], understanding programs [17], identifying objects or modules in legacy code [8, 13].

Reference [7] offers background material on lattice theory and an introduction to concept analysis. Reference [20] formalizes the notions of concept analysis and provides a proof of the fundamental theorem.

Siff and Reps [13] describe a general technique for identifying modules in legacy code. They discuss how concept analysis can identify potential modules amid tangled code using both positive and negative attributes and build a concept lattice from a program. The Tangled code occurs often in legacy systems. The authors describe a concept to be a pair of sets – a set of functions in a program and a set of properties of those functions. The properties relate the functions to data structures. The authors also present an algorithm to generating concept partitions from a concept lattice. However, the number of partitions can be exponential in the number of concepts. It may be impractical to find every possible partition of the concepts in the concept lattice. The user guidance may need to get involved in reducing the possibilities.

Snelting [16] applied concept analysis to software engineering in the NORA/RECS tool, where it was used to identify conflicts in software configuration information. Lindig and Snelting [12] and Snelting [17] applied the concept analysis to the modularization problem. In contrast with the technique presented by Siff and Reps, the concept is a pair of sets – a set of functions in a program and a set of global variables accessed by those functions.

Tonella [18] present the use of concept analysis for module identification. A modularization candidate is determined for which the variations in encapsulation and decomposition are quantified based on concept sub-partition computation.

Sahraoui et al. present an approach to support the migration of procedural software systems to object-oriented systems. The approach is very similar to ours. By using the program profiling, objects, methods, and dependencies between objects and methods are identified in a procedural code. The code corresponding to the objects and methods are transformed into object implementations. In this work, the approach is extracting objects from programs with explicit modularity. Our approach is dealing with legacy code without explicit modularity in it.

Deursen and Kuipers [8] present an approach to support identification of objects in legacy code. The approach takes the data structures as starting point for candidate objects. The authors describe the concept to be a pair of sets – a set of procedures in a program and a set of record fields accessed by those procedures. The legacy records are identified as candidate objects first. The procedures are then determined to the designated objects. However, the authors mention that legacy data structures tend to grow over time, and may contain many unrelated fields. They propose a method for restructuring the legacy data structures.

3 Overview of Formal Concept Analysis

Concept analysis provides a way to identify groupings of objects that have common attributes [20]. The mathematical foundation was laid out by G. Birkhoff in 1940 [4]. The application of concept analysis is widely used in the area of software engineering such as program understanding, restructuring of legacy code, identification of software modules, and software component retrieval. Interested readers can refer to [10] for the details of formal concept analysis and [13] for a tool implementation of formal concept analysis.

Concept analysis starts with a binary relation R between a set of objects O and a set of attributes A, hence $R \subseteq O \times A$. The triple $C = (O, A, R)$ is called a formal context. To illustrate concept analysis, we consider an example of a group of artists for painting. Users are looking for art students, art teachers, professional artists, and gallery artists. Suppose we consider five attributes: student, teacher, professional, buy, and painting. Table 1 shows which artists are considered to have which attributes. We can interpret this data in a variety of ways. For example, we might observe that art students, art instructors, professional artists, and gallery artists are related to paintings. However, only gallery artists are buying paintings.

Table 1 A formal context from the artists example

Objects	Attributes				
	student	teacher	professional	buy	painting
art students	X				X
art instructors		X			X
professional artists			X		X
gallery artists				X	X

In Table 1, the objects are art students, art instructors, professional artists, and gallery artists, the attributes are the characteristics student, teacher, professional, buy, and painting. The binary relation R is given in Table 1. For example, the tuple (art students, painting) is in R, but (art instructors, buy) is not in R. Let $X \subseteq O$ and $Y \subseteq A$. The mappings

$$\sigma(X) = \{a \in A \mid \forall o \in X : (o, a) \in R\},$$

the common attributes of X, and

$$\tau(Y) = \{o \in O \mid \forall \, a \in Y : (o, a) \in R\},$$

the common objects of Y. In the artist example, $\sigma(\{$art students, art instructors, professional artists, gallery artists$\}) = \{$painting$\}$. It indicates that art students, art instructors, professional artists, and gallery artists have a common attribute, painting. However, the gallery artists are buying paintings. So, σ ($\{$gallery artists$\}$) = $\{$buy, painting$\}$. On the other hand, $\tau(\{$painting$\}) = \{$art students, art instructors, professional artists, gallery artists$\}$. The common objects of painting are art students, art instructors, professional artists, and gallery artists. Also, $\tau(\{$buy, painting$\}) = \{$gallery artists$\}$, only gallery artists are interested in purchasing paintings. σ and τ form a Galois connection (a pair of two anti-monotone functions). That is, the mappings are anti-monotone;

$$X_1 \subseteq X_2 \Rightarrow \sigma(X_2) \subseteq \sigma(X_1)$$
$$Y_1 \subseteq Y_2 \Rightarrow \tau(Y_2) \subseteq \tau(Y_1)$$

And extensive:

$X \subseteq \tau(\sigma(X))$ and $Y \subseteq \sigma(\tau(Y))$. In the artist example, for example, $\{$art students$\}$ $\subseteq \{$art students, gallery artists$\}$ indicates $\sigma(\{$art students, gallery artists$\}) = \{$painting$\} \subseteq \sigma(\{$art students$\}) = \{$student, painting$\}$. Another example, $\{$painting$\} \subseteq$ $\{$buy, painting$\}$ indicates $\tau(\{$buy, painting$\}) = \{$gallery artists$\} \subseteq \tau(painting) =$ $\{$art students, art instructors, professional artists, gallery artists$\}$. Both $\sigma \circ \tau$ and $\tau \circ \sigma$ are closure operators: e.g., $\sigma \circ \tau(X)$ determines the biggest set of objects that have the same attributes as X. A pair (X, Y) is called a concept, if

$$Y = \sigma(X) \text{ and } X = \tau(Y)$$

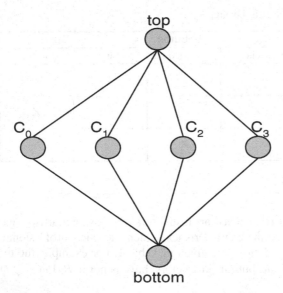

Fig. 1 The initial concept lattice for the artist example

For a concept $c = (X, Y)$, $X = \text{ext}(c)$ is called the extent and $Y = \text{int}(c)$ is called the intent of c. For example, ({art students, art instructors, professional artists, gallery artists}, {painting}) is a concept, whereas {painting} = σ(art students, art instructors, professional artists, gallery artists) and {art students, art instructors, professional artists, gallery artists} = τ({painting}). However, ({art students, gallery artists}, {painting}) is not a concept. Although σ({art students, gallery artists}) = {painting}, τ({painting}) = {art students, art instructors, professional artists, gallery artists} ≠ {art students, gallery artists}.

A concept (X_0, Y_0) is a sub-concept of concept (X_1, Y_1), denoted by $(X_0, Y_0) \sqsubseteq (X_1, Y_1)$, if $X_0 \subseteq X_1$ or equivalently, $Y_1 \subseteq Y_0$. In the example, ({gallery artists}, {buy, painting}) is a sub-concept of concept ({art students, art instructors, professional concept, gallery artists}, {painting}) whereas {gallery artists} ⊆ {art students, art instructors, professional concept, gallery artists} or equivalently, {painting} ⊆ {buy, painting}. The sub-concept relation forms a complete partial order over the set of concepts. The concept lattice for the artist example is shown in Figure 1. Each node in the initial concept lattice represents a concept. A key indicating the extent and intent of each concept is shown in Table 2.

Table 2 The extent and intent of the concepts for the artist example

top	({art students, art instructors, professional artists, gallery artists}, {painting})
c_3	({gallery artists}, {buy, painting})
c_2	({professional artists}, {professional, painting})
c_1	({art instructors}, {teacher, painting})
c_0	({art students}, {student, painting})
bottom	($\varnothing$, {student, instructor, professional, buy, painting})

The set of all concepts of a given context forms a partial order via

$$(X_1, Y_1) \leq (X_2, Y_2) \Leftrightarrow X_1 \subseteq X_2 \Leftrightarrow Y_1 \supseteq Y_2$$

G. Birkhoff discovered in 1940 that it is also a complete lattice, the concept lattice

$$\mathcal{L}(C) = \{(X, Y) \in 2^x \times 2^y \mid Y = \sigma(X) \wedge X = \tau(Y)\}$$

In this lattice, the join (infimum) of two concepts is computed by intersecting their extents:

$$(X_1, Y_1) \sqcap (X_2, Y_2) = (X_1 \cap X_2, \sigma(X_1 \cap X_2))$$

For example, ({art students}, {student, painting}) $\sqcap$ ({art instructors}, {teacher, painting}) = ($\varnothing$, σ({})) which is ($\varnothing$, {student, teacher, professional, buy, painting}). Thus, the bottom node of the concept lattice is ($\tau(\sigma(\varnothing))$, $\sigma(\varnothing)$), the concept consisting of all objects (often the empty set, as in our example) that have all the attributes in the context relation. Note that $Y_1 \cup Y_2 \subseteq \sigma(X_1 \cap X_2)$, as $X_1 \cap X_2$ has at least common attributes $Y_1 \cup Y_2$. Thus, a join describes the set of attributes common to two sets of objects. Similarly, the supremum (meet) is computed by intersecting the intents:

$$(X_1, Y_1) \sqcup (X_2, Y_2) = (\tau(Y_1 \cap Y_2), Y_1 \cap Y_2)$$

For example, ({art students}, {student, painting}) $\sqcup$ ({art instructors}, {teacher, painting}) = (τ({painting}), {painting}) which is ({art students, art instructors, professional artists, gallery artists}, {painting}). Again, $X_1 \cup X_2 \subseteq \tau(Y_1 \cap Y_2)$. Thus, a supremum describes a set of common objects that fit to two sets of attributes. In the same example, {art students, art instructors} $\subseteq \tau$({painting}) = {art students, art instructors, professional artists, gallery artists}. There are several algorithms for computing the concept lattice for a given context [13, 16]. We describe the algorithm presented in [13]. The initial step of the algorithm is to compute the bottom node of the concept lattice. The next step is to compute atomic concepts – the concepts formed from the singleton sets of objects, i.e., ($\tau(\sigma(\{x\}))$, $\sigma(\{x\})$) for each object x. Computation of the atomic concepts for the artist example is shown in Figure 2.

$\tau(\sigma(\{\text{art students}\})) = \tau(\{\text{student, painting}\}) = \{\text{art students}\}$

$\tau(\sigma(\{\text{art instructors}\})) = \tau(\{\text{teacher, painting}\}) = \{\text{art instructors}\}$

$\tau(\sigma(\{\text{professional artists}\})) = \tau(\{\text{professional, painting}\}) = \{\text{professional artists}\}$

$\tau(\sigma(\{\text{gallery artists}\})) = \tau(\{\text{buy, painting}\}) = \{\text{gallery artists}\}$

Fig. 2 Computing atomic concepts in the artist example

We have briefly overviewed the concepts of formal concept analysis. A web site containing excellent resource for formal concept analysis can be referenced in [9]. This web site delivers the information about the tools of formal concept analysis and the conferences on formal concept analysis.

4 Overview of the Approach

Non-object oriented code does not contain an explicit representation of objects. It only contains global variables, data structures, and modules. In this section, we propose an approach using formal concept analysis to identify objects in procedural programs, particularly in COBOL programs. The goal of this research is to help the creation of objects from procedural programs by grouping records and the associated statements. The software architecture of the implementation of this approach is an instance of pipeline software architecture [15] where the output of each phase is the input to the next phase. The processing steps are illustrated in Figure 3.

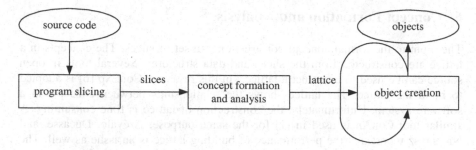

Fig. 3 The steps of the process

Each step is briefly explained as follows:

- Program slicing. An object is composed of a set of data and a set of operations that manipulates the data. We start with a data structure and extract the code corresponding to the data structure for object identification and creation. Many tools [2] can be used to understand the programs for this process.
- Concept formation and analysis. Some languages like COBOL do not support objects explicitly. This step is to identify objects in the slices using formal concept analysis. A concept is identified in terms of a set of slices and data. The concepts represented in groupings in a lattice are generated by a formal analysis tool, namely ConExp [6]. Unsuitable groupings are filtered out and the rest of groupings are the candidates for the final generation of the objects.
- Object creation: this step is to migrate the code in the slices into object implementations.

5 Program Slicing

The input to this step is a software system. Several recommendations are listed here before the process. The code has better no goto statements in it. This is an important issue because many commercial tools applying program slicing might assume the code has no goto statements. It would be also better to have dead code removed from the code. However, this won't impact the results because theoretically removal of dead code should be subsumed by the removal of clones of concepts.

The concept of program slicing can be found in the Weiser's pioneer work [19]. A slice is considered as a fragment of code in a program. We begin to identify interested data structures as a starting point. For each data structure, we then collect the statements that operate on the data structure. A data structure can be composed of a set of data items. The statements in a slice might be collected across the program and not necessarily to be located in a module, procedure, and function. In [5], Chiang presents the details of a program slicing technique to the extraction of the code from a program. The results of this step are a set of program slices relevant to the interested data structures. The data structures and the associated code in the slices might form the candidate objects.

6 Concept Formation and Analysis

The input to the formal concept construction is a set of slices. The concepts in a lattice are constructed from the slices and data structures. Several tools in open source can be used to construct a lattice. In this research, ConExp [6] is adopted to build the lattice. The lattice shows the relationships between concepts in a software system. Unfortunately, the construction of lattice is time consuming. A similar tool ConAn is used in [1] for the same purpose. Arévalo, Ducasse, and Nierstrasz report that the performance of building lattice is an issue as well. The time complexity is $O(n^2)$ where n is the number of concepts calculated.

Once the lattice is built, each grouping in the lattice constitutes a candidate object. But not all the groupings are meaningful, a filtering process is needed to filter irrelevant ones. For example, the top and bottom groupings should be removed from the candidate object list. Disjoint groupings should also be removed from the list because the interesting characteristic of the approach is to find reusable objects sharing common codes. In [14], a set of heuristics is suggested to filter out the set of candidate objects.

At this stage of the project, this step has not been automated yet. Our experiences show that the process is subjective and requires experienced software engineers with the domain knowledge of the application to do the task. The step is also iterative. A software engineer needs to repeat the analysis until a final list of candidate objects is made.

7 Object Creation

The lattice helps software engineers to identify objects by grouping data structures and code in slices that might qualify for objects. This step is to move the code from the slices into objects involving aggregation and generalization. Aggregation moves the code from different slices to the objects. In addition, the code in the methods might be modified so it can be invoked by all the programs in the application.

8 Lessons Learned

What lessons have we learned from this research? It is difficult to determine which data structures are needed to extract slices for objects. The adequate choices of data structures of interest determine the identification of slices. Different names of data structures may indicate the same one. Having the knowledge of the programs may help ease the task.

The code extraction corresponding to the data structures is also a time-consuming task. The results will produce an initial set of slices. Some slices may be subsets of the other slices. Some slices might not be usable for creating objects. Therefore, at this stage of the research, the approach requires software engineers to manually filter the slices for the best ones. It would be better to have experienced software engineers to do this task. The rule of thumbs is that the final chosen candidates must generate the maximum benefit from the perspective of software reuse. The filtering process is also very subjective because it depends on the knowledge and experience of the software engineers.

The interpretation of a grouping in a lattice for a concept is a frustrating and time-consuming task for software engineers. Software engineers are not necessarily to be experts of formal concept analysis. It is a subjective for software engineers to identify concepts. The performance of formal concept analysis algorithm is another issue of using this approach. Large cases may require several hours to construct a lattice.

Different programming styles and languages impact the approach to object identification in a legacy code. It is impossible to just adopt an approach to accomplish the task. A legacy application is usually developed in different programming languages and database systems. Thus, a combination of approaches might be needed.

9 Summary

In this paper, we presented an approach to identify the objects in legacy procedural programs. The approach is very different from existing approaches. Our approach is not only to identify the objects in procedural code but also to migrate the code from the procedural code into objects where the corresponding code may be dispersed in different locations across the entire system. At this stage of the research, the approach requires software engineers to select the best candidates for the objects. Future work may include the development of a tool to ease the process semi-automatically.

References

1. Arévalo, G., Ducasse, S., Nierstrasz, O.: Lessons learned in applying formal concept analysis to reverse engineering. In: Ganter, B., Godin, R. (eds.) ICFCA 2005. LNCS (LNAI), vol. 3403, pp. 95–112. Springer, Heidelberg (2005)
2. Arnold, R.: Software Reengineering. IEEE Computer Society Press, Los Alamitos (1993)
3. Belady, L.A., Lehman, N.M.: A model of large program development. IBM Systems Journal 15(3), 225–252 (1976)
4. Birkhoff, G.: Lattice Theory, 1st edn. American Mathematical Society (1940)
5. Chiang, C.-C.: Extracting business rules from legacy systems into reusable components. In: SoSE, pp. 350–355. IEEE, Los Alamitos (2006)
6. ConExp., http://conexp.sourceforge.net/users/index.html
7. Davey, B.A., Priestley, H.A.: Introduction to Lattices and Order. Cambridge University Press, Cambridge (1990)
8. Deursen, A.V., Kuipers, T.: Identifying objects using cluster and concept analysis. In: ICSE, pp. 246–255. ACM Press, New York (1999)
9. Formal Concept Analysis, http://www.upriss.org.uk/fca/fca.html
10. Ganter, B., Wille, R.: Formal Concept Analysis: Mathematical Foundations. Springer, Heidelberg (1999)

11. Krone, M., Snelting, G.: One the inference of configuration structures from source code. In: ICSE, pp. 49–57. IEEE, Los Alamitos (1994)
12. Lindig, C., Snelting, G.: Assessing modular structure of legacy code based on mathematical concept analysis. In: ICSE, pp. 349–359. IEEE, Los Alamitos (1997)
13. Siff, M., Reps, T.: Identifying modules via concept analysis. IEEE Transactions on Software Engineering 25(6), 749–768 (1999)
14. Sahraoui, H., Lounis, H., Melo, W., Mili, H.: A concept formation based approach to object identification in procedural code. Automated Software Engineering 6, 387–410 (1999)
15. Shaw, M., Garlan, D.: Software Architecture. Prentice-Hall, Englewood Cliffs (1996)
16. Snelting, G.: Reengineering of configurations based on mathematical concept analysis. ACM Transactions on Software Engineering and Methodology 5(2), 146–189 (1996)
17. Snelting, G.: Concept analysis – a new framework for program understanding. In: Flanagan, C., Zeller, A. (eds.) Sigplan/Sigsoft Paste, pp. 1–10. ACM Press, New York (1998)
18. Tonella, P.: Concept analysis for module restructuring. IEEE Transactions on Software Engineering 27(4), 351–363 (2001)
19. Weiser, M.: Program slicing. IEEE Transactions on Software Engineering SE-10(4), 352–357 (1984)
20. Wille, R.: Restructuring lattice theory: an approach based on hierarchies of concepts. In: Rival, I. (ed.) Ordered sets, pp. 445–470. NATO Advanced Study Institute (1981)

Multiple Factors Based Qualitative Simulation for Flood Analysis

Tokuro Matsuo and Norihiko Hatano

1 Introduction

Recently, there are a lot of researches concerned with natural disaster simulation by using data gathering useful information from sensors and climate management systems. However, most of simulations are based on high specs computers since the simulation is conducted with numerical and quantitative method[1][2][4][7]. By using huge data such as climate, temperature, and other physical status about degree of altitude, the simulation gives a rigorous result of analysis even though the situation and circumstance is complex like property of the atmosphere. Further, they are also developed for specialist but naive users and novices.

The advantages of qualitative reasoning in education are as follows. Student knowledge is formed and developed through learning conceptual foundations. If there are any mechanisms in the (dynamic) system, the user can understand these mechanisms using qualitative methods. Generally, beginners(students) also understand dynamic systems through qualitative principles, rather than through mathematical formula. In that case, when senior learners, such as university students, learn liquid and flood dynamics using current system, users may be able to understand the mechanism easily. Letfs consider user grade is less than high school student. It is difficult for them to learn and understand using economical formula. In our study, we developed our approach in least formula and took learning by non-specialist users into consideration.

In existing studies, most of simulations in dynamics are based on numerical calculations. However, naive and novice learner, such as elementary school

Tokuro Matsuo and Norihiko Hatano
Graduate School of Science and Engineering,
Yamagata University.
Jonan 4-3-16, Yonezawa,
Yamagata, 992-8510 Japan
e-mail: tokuro@tokuro.net
http://www.tokuro.net

R. Lee, N. Ishii (Eds.): Software Engineering, Artificial Intelligence, SCI 209, pp. 179–189.
springerlink.com

students and junior high school students, cannot understand numerical calculation because numerical-based simulation generally employs certain complicated formula. Users cannot know the learning of the formula, such as differential equations. We employ qualitative methods in place of quantitative methods. Our qualitative simulation model is constructed as causal graph model. The model consists of nodes and arcs connected to nodes.

The rest of the paper is organized as follows. Section 2 outlines the environment of the system and qualitative reasoning. In Section 3, we propose and define the model of qualitative simulation that includes qualitative status values and changing trend of influences in the natural system. In Section 4, we show a simulation process. Then, in Section 5, some examples of qualitative simulation for flood analysis are shown. After that, in Section 6, we discuss precondition and data used in the simulation. Finally, in Section 7 we provide some final remarks.

2 Outline

2.1 System Architecture

In this paper, we propose a natural disaster forecast support system with qualitative simulation. We assume, in this paper, that the system has already gathered information about climate, geography, and rainfall from multiple systems and sensors. Figure 1 shows the concept in which the system gathers data and information. As actual process, first, a system manager inputs

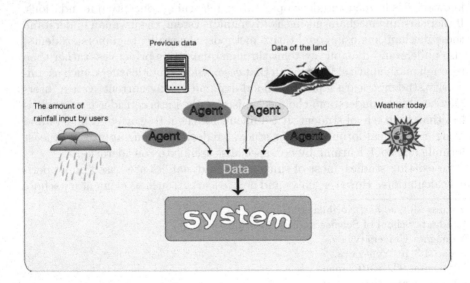

Fig. 1 Outline of the system

geographical and general rules. General rule is knowledge like the Newton's law and is never changed and modified. Then, users input local geographical information and history of observational data.

2.2 Qualitative Reasoning

Advantage of qualitative reasoning is compare qualitative model with quantitative model, it models based on the acknowledgment process of man[3][5][6]. Therefore, even if it is neither a scholar it is nor a specialist, it is easy to understand the process and the result of the idea and the behavior of the model. In fact, the whole image (in this case, natural phenomenon) is grasp to entire picture in broad-brush shape, and it is possible to understand easy.

Grounds run a simulation with the use of qualitative reasoning, run a system that natural disasters, on dynamic behavior by ordinary not differential equation in behalf of quantitative construal, based on the qualitative differential equation dynamic behavior make a deduction perccive dynamic behavior through qualitative construal. Accordingly, we simplify the process and calculate. Using this factor can drastically curtail the process.

Generally, we can enough calculate the qualitative reasoning, when we use only home computers.

3 System Modeling

3.1 Modeling of Factor and Natural Phenomenon

We consider a modeling based on the natural phenomenon like flood. The system runs a simulation based on models based on some factors of natural phenomena. That factor is amount, capability dewaters from field, and remaining water dosage. Natural phenomena change based on some factors, which relate each other.

3.2 Feature of Modeling

When the flood simulation is conducted, the geographical space and span should be defined just a location where the user wants to know the condition of flood. The span is defined based on the width and number of fields. The general geographical condition is input by a system manager or an end user. We assume width of each field is same physical space and dimensions. We also assume each field is square.

There are multiple factors regarding a condition of each field. In our system, user inputs the amount of rainfall when the data do not come from rain sensor. In this paper, to analyze the general case of flood, we assume the amount of rain per minute is based on at random. After user inputs total amount of rain, the simulator allocates multiply with divided time span.

Here, we consider the drainage-capability. End user inputs the rate of drainage capability to the system. Even though the flood reduces due to high performance of drainage capability, we assume the capability decreases with drainage since fields and grounds are saturated. However, we assume the drainage-capability does not become zero. When there are no water to gobble up, the capability increases. The highest value of capability does not go up more than the value in which user inputs.

Finally, regarding the remaining water at the field, we employ the qualitative feature of it. Concretely, we assume there are five levels of remaining water in the field. Each condition has the features with its changing rate of water as follows.

- When the ability of rainfall is more than drainage capability and remaining water is more than drainage capability per minute, we set the condition as "++".
- When the ability of rainfall is more than drainage capability and remaining water is more/less than drainage capability per minute, we set the condition as "+".
- When the ability of rainfall is same as drainage capability and remaining water is more/less than drainage capability per minute, we set the condition as "0".
- When the ability of rainfall is less than drainage capability and remaining water is more/less than drainage capability per minute, we set the condition as "-".
- When the ability of rainfall is less than drainage capability and remaining water is less than drainage capability per minute, we set the condition as "- -".

4   Definition in Model

4.1   Flood-Simulator in Case of Single Field

We describe definition about flood-simulator in case of single field.

Precondition

Prerequisites of our simulator simulate the flood at the time of the heavy rain. When the river is assumed as the simulated field, the normal level of water is set as landmark for the simulation. We assume that the condition of a field is divided with one minute. We also assume the remaining water of the field is less than zero.

We define the changing state of remaining water as "-1", "0", and "1" since the computer calculate the condition based on qualitative simulation. Changing of qualitative value becomes "+2" on specific condition. Width can have be in change of value. The drainage capability is reduced when the work continues. When the sum of the amount of rainfall and remaining water is zero, the drainage capability is equal to or less than the ability in the previous step.

Definition

We define about the qualitative calculation method to conduct simulation. The definition of qualitative condition is consists of condition of field and changing trend of water and field's characteristics. Here, we give definitions of qualitative values of condition.

Qualitative status value of remaining water

- I : The qualitative value doesn't reach a value that is lower than the value before one minute .
- J : The qualitative value doesn't reach a value that is lower or higher than the value before one minute.
- K : The qualitative value doesn't reach a value that is higher than the value before one minute.

Next we show the changing tendency of state of remaining water.

Changing tendency of a state as qualitative value

- H : The qualitative value increases more than the value before one minute.
- Z : The qualitative value keeps as the same value before one minute.
- L : The qualitative value decreases from the value before one minute.

Next, logically, we remove the situation where the qualitative definition is paradox.

For example, when the combination of (I, H), it is valid with the above qualitative definition of state and changing trend. Thus, the combination of

Table 1 Exception of Absurdity of qualitative value

	I	J	K
H	(I,H)=+	(J,H)= Empty	(K,H)= Empty
Z	(I,Z)=0	(J,Z)=0	(K,Z)=0
L	(I,L)=Empty	(J,L)=Empty	(K,L)=-

them is described as "+". Second, we consider the combination of (I, Z). When the combination of (I, H), it is also valid with the above qualitative definition of state and changing trend. Thus, the combination of them is described as "0". Third, the combination of (I, L) is not valid since there is paradox in relationship between "I" and "L". Thus, the combination of each value is shown as the table 1.

Process of Simulation

We explain the process of simulation as Figure 2. First the function for calculation is defined. Second, User inputs the amount of rainfall and drainage. Then, The rainfall is allocated to each divided time step. Forth, the system compare the amount of rainfall and the drainage capability based on the qualitative value and rules. After that, the amount of rainfall is re-compared with the drainage capability. The system continues the above process. When a certain time step defined by user comes, the simulation is finished.

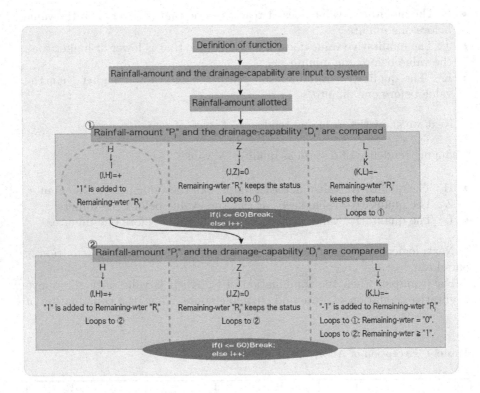

Fig. 2 Process of simulation with single field

5   Result of Simulation

The result of the simulation is shown in this section. The simulation indicates the flood of the fields. The condition of simulation is as follows.

- Number of field: 1
- The amount of rainfall: $10[mm/h]$
- Drainage capability: $40[mm/m(MAX)]$

We give a situation of changeable amount of rainfall and drainage in Figure 3. The horizontal axis shows the time and left vertical axis shows the amount of rainfall and drainage. Right side of vertical axis shows the qualitative values. Small broken line represents the ability of drainage and bid broken line represents the remaining water at the field. The bar graph shows the amount of rainfall. Concretely, drainage ability reduces since the drainage is continued. After twenty-seven minute, the amount of rainfall is more than the ability of drainage. Thus, the drainage changes. Further, after thirty-three minute, the remaining water is more than drainage ability. In this case, it rains on the field, the remaining water of the field increases.

Here, we consider the improvement of drainage ability. We give a condition as follows.

- Number of field: 1
- The amount of rainfall: $1[mm/h]$
- Drainage capability: $12[mm/m(MAX)]$

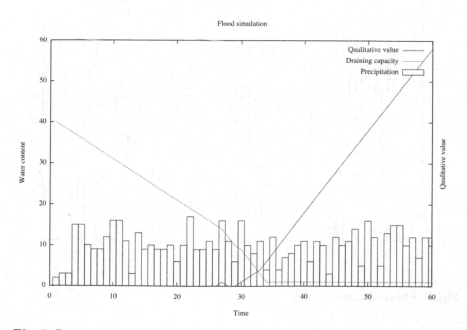

Fig. 3 Drainage of rainfall-water

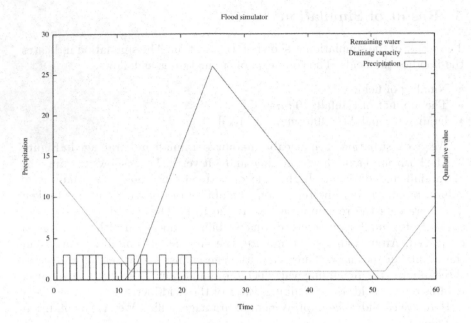

Fig. 4 Improvement of drainage ability

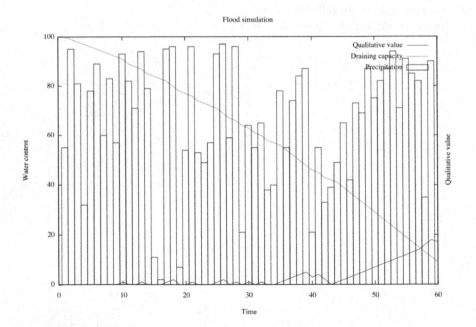

Fig. 5 Flood simulation

Figure 4 shows the situation where the ability of drainage gradually improves. The value of drainage capability does not become less than its initial value. Concretely, after twenty-five minutes, the amount of drainage and remaining water is same as Figure 3. When the remaining water at the field become zero, the ability of drainage recovers.

By using the above condition, we conduct the simulation of water flood.

- Number of field: 1
- The amount of rainfall: $65[mm/h]$
- Drainage capability: $100[mm/m(MAX)]$

Figure 5 shows the situation, where the amount of water is increases when the ability of drainage decreases. For example, the value of drainage never increases since the amount of rainfall is very huge. The amount of rainfall is more than the ability of drainage. The remaining water increases more and more without reducing. Thus, the water flood occurs.

6   Discussion

6.1   Flood-Simulator in Case of Multiple Field

We show the definitions and rules about the multiple fields.

Precondition

A basic precondition is the same as a single field. However, it different types of definition about the just diffusion of water. We consider two types of definition, such as, water flow from other field and no water flow. We do not consider the height of field.

Definition

When there are multiple fields such as an actual geographical situation, the simulation model develops more complicated since the relationship between each field is increases. Even though the model is complex and computational costs increase, the basic idea of simulation is same as the single field simulation. In this paper, we handle the situation of line like river. The field makes a line like nose to tail. In this simulation, we arrange fields in line like $1 \times n$. We take the transmission speed in the point of width of field into consideration. If the drainage-capability is a low place, "1mm" is sure to remain in the spreading water by the surface condition.

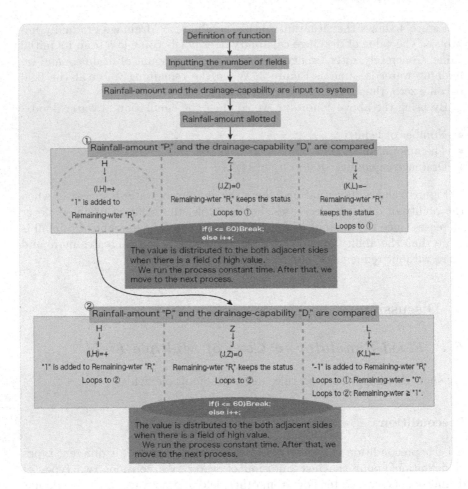

Fig. 6 Process of simulation with multiple fields

Flow of Data

We consider about the process of simulation.

Figure 6 is a process of simulation where there are multiple fields. First, the user defines the function of simulation. Then, he/she inputs the number of field, amount of rainfall and drainage ability. After that, the amount of rainfall is compared with drainage ability with each step. This process is continued until the time in which the user set up. In this case, the graph model becomes very complex.

7 Conclusion

In this paper, we proposed the qulitative simulation based flood analysis and forecast system to make students understand without differential equations.

The system gives the simulation result qualitatively. In qualitative simulation, qualitative model has multiple factors, such as the amount of rainfall, drainage, and remaining water. Each factor also has qualitative value with each time step. Thus, the dynamical phenomena can be expressed conceptually with visual graph and can be understood easily by learners who has not known the numerical/dfferential equation.

References

1. Agell, N., Aguado, C.J.: A hybrid qualitative-quantitative classification technique applied to aid marketing decisions. In: The proceedings of 11th International Workshop on Qualitative Reasoning (2001)
2. Bredeweg, B., Winkels, R.: Qualitative models in interactive learning environments. Interactive Learning Environments 5 (1998)
3. Chen, S.A., Wang, J., Yang, C.S.: Constructing internet futures exchange for teach-ing derivatives trading in financial markets. In: The proceedings of International Conference on Computers in Education, vol. 2, pp. 1392–1395 (2002)
4. Forbus, K.D.: Helping children become qualitative modelers. Journal of the Japanese Society for Artificial Intelligence 17(4) (2002)
5. Hata, S., Ohkawa, T., Komoda, N.: Backward simulation method in qualitative simulation. IEEJ Transactions on Electronics, Information and Systems, Institute of Electrical Engineers of Japan 115-C(11) (1995)
6. Kuipers, B.: Qualitative Reasoning. The MIT Press, Cambridge (1994)
7. Matsuo, T., Ito, T., Shintani, T.: A qualitative/quantitative methods-based e-learning support system in economic education. In: The 19th National Conference on Artificial Intelligence (AAAI 2004) (2004)

Layering MDA: Applying Transparent Layers of Knowledge to Platform Independent Models

Oliver Strong, Chia-Chu Chiang, Haeng-Kon Kim, Byeongdo Kang, and Roger Lee

Abstract. Model Driven Architecture is a system that holds promise for software engineers, but intimidates them also. This is due to the fact that in order to fully take advantage of MDA engineers must be fluent in UML as well as the languages used to translate their PIM's into executable. Though UML has been in existence since the mid 1990's its use for communication among people is eased by shared understanding and inferences, these two things are not available amongst computer interpretation units. That means that in order to satisfy automated translation units SE's must generate PIM's and in some cases manually manipulate PSM's that are obfuscated in the eyes of not so well versed individuals. That is why we propose a system of dynamic markups that can provide details that would otherwise be incomprehensible, by people outside the circle of those highly educated in UML and MDA. These markups are non-translatable and can be specialized to certain groups of shareholders. Finally, a small demonstration of this system is performed in order to allow for a fuller understanding of the process and concept, while clarifying the benefits, also to set the grounds, by which it can be empirically proven.

Keywords: Model Driven Architecture, Markup, Platform Independent Model, Ambiguity.

Oliver Strong
Computer Science Department Central Michigan University, United States
e-mail: stron1om@cmich.edu

Chia-Chu Chiang
Computer Science Department University of Arkansas-Little Rock, United States
e-mail: cxchiang@ualr.edu

Haeng-Kon Kim
Department of Computer and Communications Engineering
Catholic University of Daegu, Korea
e-mail: hangkon@cu.ac.kr

Byeongdo Kang
School of Computer & Information Technology Daegu University, Korea
e-mail: bdkang@daegu.ac.kr

Roger Lee
Software Engineering & Information Technology Institute
Central Michigan University, United States
e-mail: lee1ry@cmich.edu

R. Lee, N. Ishii (Eds.): Software Engineering, Artificial Intelligence, SCI 209, pp. 191–199.
springerlink.com   © Springer-Verlag Berlin Heidelberg 2009

1 Introduction

Model Driven Architecture is a process that allows for a discrete and absolute separation of platform details from the chosen platform of implementation. MDA, then by the merits of this separation is capable of increasing both the longevity and conciseness of software entities. While this is commonly understood and accepted it the implementation of this schema that causes unrest, by those that wish to study it and by those that wish to utilize it.

Those that wish to study it find that there is an important breakdown in methodologies by which MDA is executed. These breakdowns leave MDA practitioners in two loosely organized camps. McNeile terms these camps "Elaborationist" and "Translationist" [7].

The first category, the elaborationist find themselves elaborating by hand the code that is produced from one of many translation entities they may use. This elaboration is required since many of the target platforms differ greatly such as with enterprise software systems and distributed applications [5]. Generating up to 80% of the code through several translation units intermediary PSM's are required to be human understandable for at least a portion of the development process(Haywood).

The other side of the coin has a group of software engineers that require 100% of the executable to be generated via translating the PIM. This group focuses on real time and embedded systems which change platforms rapidly and have a much shorter life span [7].

Both methodologies suffer however from the same dilemma. Translationist find their roots in the non-automated generation of software, particularly the Object Oriented Programming, this being high level diagrams to specify a system (Object Oriented Design) to the implementation phase. The difference between OOP and MDA lies in the fact that MDA has tools to generate code instead of implementing it manually.

Elaborationist find that in order to be the most efficient in their work they must produce code that is translatable in one step with little to no interaction by the developers. This dependence on machine interpretable translations creates a level of UML that is executable and purposeful, but highly complex and unintelligible to any, but the highly trained.

Table 1 Translation v. Elaboration

Approach	Field	Number of PSMs	Point at which UML becomes difficult to understand
Translation	Embedded/ Realtime systems	1	PIM
Elaboration	Enterprise Systems	Many	PSM

1.2 Errors in Communication

It has now been stated in the manner that MDA techniques make project facts incomprehensible. It now will be shown the manners that these incomprehension arise from the three main areas;

1. Excess PIM level abstraction, which is necessary for compiling PIMs to PSMs.
2. Proliferation of PSM level documents.
3. Overall complex language syntax with a high learning curve.

A Platform Independent Model by definition contains no platform details merely the ideas or constructs of the system not the means by which the system(s) will be implemented [8]. It is demonstrable how a platform model may cause issues that are not resolved in the CIM, Computation Independent Model. The CIM maintains details of the system without consideration to implementation (MDA specification). For the reason of scope CIM's will not be dealt with in this document.

1.3 Ambiguity of Platform

The concurrent generation of platform models and translation units cause a breakdown in communication between stakeholders and developers. As the project progresses it becomes necessary to communicate between these parties without much overhead. The UML being understood at least in part by all parties, but the diversity of the intellectual background and field of knowledge between parties can be cumbersome to convey knowledge.

The differences in platform can be quite tremendous, though appear to be non-existent when you solely viewed from a platform independent aspect. Such, miniscule differences in the PIM can be very large when there is a disparity between platforms.

This difference can be very easily explained and understood in subsequent forms and methods of design, but not readily available in the PIM. That is why the system proposed allows for easy configuration of expendable executables that can impose permutations upon the PIM that are transparent to all, but the stakeholder that inflicted the changes.

Using such techniques to impose data upon a unit that will be computer rendered into a different format is not a new concept. It has been in use by drafts people utilizing Computer Aided Design software since its [9].

2 Method

A PIM by definition contains no platform implementation details and furthermore can be translated to a platform specific model in some method as deemed most appropriate by the developers, though arguably this translation is intended to be performed by the use of translation tools that are executed on computers in order to fully benefit from MDA over a process such as OOD where models can be used, but there is no standardized mechanism of translation[4]. Though the most appealing benefit of MDA is the fact that not linking platform detail to functional concepts allow for the use of PIM's over and over by applying new translation mappings [3].

Translation units on average are large and costly being that are designed mainly for proprietary use MDA tool vendor's offerings, though it has been argued that as MDA becomes more popular so will the translation units, and as such laws of supply and demand offer so the price will fall[4][6]. Yet, if you merely have a repository of PIMs, PSMs and finally some executable you have a set of very complex and in some cases lacking models. Using strict PIMs to get the best translation using the automated translation technique may make it complicated to share, calculate or understand the implications of certain functionality on specific hardware.

Accounting for the aforementioned points leads to the following main points of interest:

Question: Can simple platform details be added to PIM's without the overhead of complex translation units?
Question: Would dynamic addition of 'shallow' platform factoids make PIM's more useful?
Hypothesis: Platform details may be added or fetched in any method desired and transparent layers of knowledge on top of the PIM will prove to be beneficial.

2.1 Test Case

In order to best extrapolate the problem at hand a test case has been devised that proliferates the issues MDA has with structural and contextual ambiguity. The ambiguity being found in this test case is obvious, but will be called out and addressed and then followed up with a working model of the solution.

Figure 1 is a portion of a PIM detailing a system used to loopback what will be referred to as Media X. Media X is located in some location that the application is aware of. An unknown number of Media X items will be placed in a queue until all items are queued, simultaneously loopback devices of a constant, but unknown size will be propagated, capable of writing Media X items as they are played to an internal capture interface. Upon the emptying of the queue some time will have passed and it is the purpose of test case to approximate how much time will be necessary to complete the copying on various platforms with varying sized Media X collections.

2.2 Understanding the Problem

The diagram presented leaves much to ambiguity even though understanding the trivial aspects are simple and most likely do not need to be enumerated in real world scenarios, but make a good case point for academic purposes.

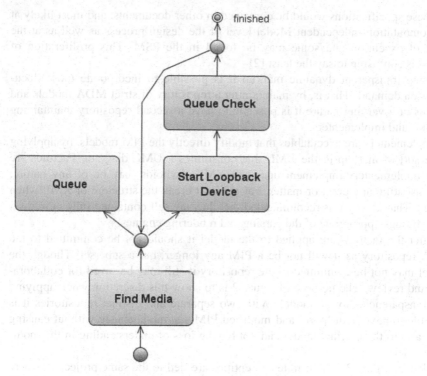

Fig. 1 Loopback recording device process diagram

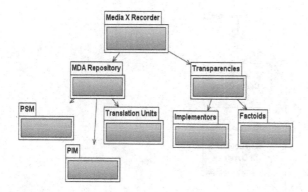

Fig. 2 Hierarchy of recording system development files

Simply speaking, it is impossible to tell if this tool will meet any requirements from this diagram. And in reality there would be many complimentary documents to follow, but none (at the PIM level) would specify the size of a recording entity, the memory of the target system, process limits of the target system.

These specifications would be contained in other documents, and most likely at the computation independent Model level of the design process as well as at the point of execution and some may be found in the PSM. This proliferation of details is confusing to say the least [2].

Using transparent dynamic markup it is possible to incorporate those documents on demand. That is, by maintaining a repository of strict MDA models and entities of a varying nature it is possible to have a second repository maintaining factoids and implementers.

Implementers are executables that modify directly the PIM models, by applying some sort of markup to the XML that constitutes a UML diagram. Factoids are what implementers implement upon the XML. Factoids may be of any nature, demonstrating any point or matter, but may not break the structure of XML when parsed. That is why it is recommended that they are self contained fully closed set of XML tags appropriate of the parsing and rendering engines.

After the factoids are applied to the model it should not be committed to the MDA repository as it will not be a PIM any longer, but a superset. Though the model may not be committed to the repository it still may be saved for collaboration and review. The purpose of figure 2 is to show this discretion when applying the transparencies to the model. With two separate and distinct repositories it is possible to have both 'pure' and modified PIM's simultaneously without causing any harm to the original model and not having loss of understanding in the modified model.

Also, as figure 2 demonstrates all entities are tied to the same project so where one is present the other is also. This is important in the fact that you will not have

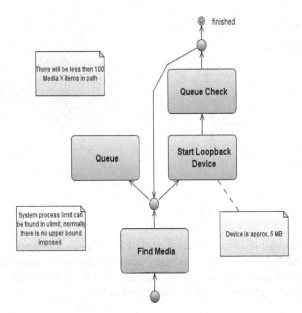

Fig. 3 Figure 1 with transparencies applied

to worry about not having the shallow transparencies to guide your understanding of the purer PIM models as they will always be present together, if not present in a combined form.

Back to the Media X example at hand. It is not available from the diagram in figure 1 many factors concerning timing and system load of the example problem. Adding these details using transparencies is remarkably simple. Without the need to demonstrate the code a couple of sample markups can be produced.

Figure 3 and figure 4 are both the result of a separate implementer. Each implementer performs a unique and specific task. In figure 3 it gives information as to the overhead and time required to translate a number of entities per system parameters. In figure 4 overhead and functional understanding of the Media X acquisition and processing may be gleaned.

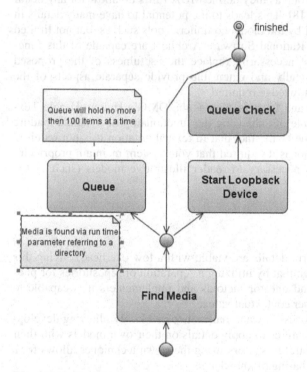

Fig. 4 Figure 1 with a separate set of transparencies showing system versatility

2.3 Algorithm

The algorithm at has is very simple and is as follows:

1. Create PIM.
2. Determine factoids place in repository. *
3. Create Implementers.
4. Apply implementers as necessary.
   * 2 and 3 are transposable.

The elegance of this algorithm is that it is totally optional with only the parties that wish to partake assuming the overhead of utilizing the method. With such low cost it requires very little reason not to implement such a method of shallow transparent modeling.

Like MDA itself this technique is merely a set of steps and not a confinement of proprietary languages and tools. Your most convenient languages and already existing project portions may be utilized along with many other languages and techniques.

3 Related Works

The Object Management Group as they define MDA PIMs do allow for any detail that is not platform specific [8]. This leads to the potential to have many details in the PIM that can be ignored by the model compiler. Tools such as, but not limited to Netbeans, OptimalJ and Rational Software Architect are capable of this functionality though it does not necessarily replace the usefulness of the proposed system as it cannot dynamically and extensible provide separate aspects of the same model based on the knowledge required.

Furthermore, techniques and formats such as the OMG's Human Usable Textual Notation, may be capable of enhancing the functionality of dynamic loading of transparencies. This is due to the fact that in textual notation it is not required that you understand UML nor is it required that your system maintain proprietary software that is sometimes necessary to render illustrative models (such as the ones used in this test case).

4 Conclusions

Dynamic transparent platform details are viable with a low overhead. Saying this means that it has been shown that by utilizing a separation of repositories for platform independent details and one for factoids and implementers it is capable to produce diagrams for a higher contextual value.

The simplicity of this process is what makes it so viable. By allowing developers and stakeholders of all natures to apply details on their own models with their additions submitted to a central repository using their own techniques allows for a greater and more dynamic sharing of knowledge.

Furthermore the system is self propagating and sustaining it. That means that you provide the factoids and implementers and there is no upkeep they merely reside in the repository until you wish to use them. If at anytime they become unnecessary or no longer function as intended they can be fixed or ignored as they are not crucial.

In part it has been shown that by using simple and strait forward markup of otherwise complex UML the learning curve and communication gap have been breached. UML has been made more accessible to all parties as well as the platform details have been revealed.

5 Future Work

Future work could make shallow details an intricate and invaluable part of MDA. As it stands the philosophical and underlying ground work has been laid to facilitate a mechanism for providing a consistent and attractive interface to what has been demonstrated in the body of this work. As part of an MDA toolset or merely a third party extension that would allow for an easy and quick application of transparent layers this new technology could be brought to everyone from the engineers to the stakeholders that wish to review their investments without undertaking an extensive and costly training in advance software engineering techniques.

Once an appealing and simple interface has been contrived it could be determined if this system could scale as proposed. Though it has been shown that these details can be applied and do have a benefit to practitioners, it is not possible from this work to derive if it would be adopted assuming availability.

References

1. Asadi, M., Ramsin, R.: MDA-Based Methodologies: An Analytical Survey. In: Schieferdecker, I., Hartman, A. (eds.) ECMDA-FA 2008. LNCS, vol. 5095, pp. 419–431. Springer, Heidelberg (2008)
2. Bettin, J., Boas, G.v.: Soft Metaware (2002, 10 22), http://www.softmetaware.com (retrieved, 10 03, 2008)
3. Capitaine Scott, M.-P.G.: Composition rules for PIM reuse (2004)
4. CodeFutures . UML Tools: Java Code Generation versus UML Tools. (2008), http://www.codefutures.com/uml-tools/ (retrieved, 01 02 2009)
5. Haywood, D.: MDA: Nice Idea, Shame About the.. (2004 05), http://www.theserverside.com (retrieved 12 03, 2008)
6. Kontio, M.: Architectural manifesto (2005, 08 17), http://www.ibm.com/developerworks/wireless/library/wi-arch17.html (retrieved, 9 3, 2008)
7. McNeile, A.: MDA: The Vision with a Hole (2003), http://www.metamaxim.com (retrieved, 11 13, 2008)
8. Object Management Group. MDA (08, Janurary 2001), http://www.omg.org/mda/ (retrieved, 12 12, 2008)
9. Virginia Technological University. Facilities at VT (08, 2003), http://www.facilities.vt.edu (retrieved 12 06, 2008)

Semantics Based Collaborative Filtering

Jae-won Lee, Kwang-Hyun Nam, and Sang-goo Lee

Abstract. Collaborative filtering is one of the most successful and popular methodologies in recommendation systems. However, the traditional collaborative filtering has some limitations such as the item sparsity and cold start problem. In this paper, we propose a new methodology for solving the item sparsity problem by mapping users and items to a domain ontology. Our method uses a semantic match with the domain ontology, while the traditional collaborative filtering uses an exact match to find similar users. The results of several experiments show that our method is more precise than the traditional collaborative filtering methodologies.

1 Introduction

The number of items (music, movie, web documents, and so on) available on the Web has grown continuously since the advent of the Web. As a result of the growth, it becomes increasingly difficult to find items that are relevant to users' preferences. In order to provide adequate items to the users, two types of techniques are widely used; Recommendation and Search.

Recommendation systems look for the users' needs more actively, whereas *search systems* receive the needs passively. That is, while most *search systems* ask users to submit information needs with queries, *recommendation systems* do not. Requiring users to submit queries often has problems. For example, if a user does not know keywords of item exactly, he/she may hesitate what queries have to be submitted. Moreover, if a user submits wrong queries that are not relevant to items, he/she may be dissatisfied with search results. To solve these problems, recommendation systems are proposed. Collaborative Filtering (CF) is one of the most successful and popular methodologies in recommendation systems. The reasons of its success are as follows. First, it is possible to reflect users' preferences through ratings of items. Second, it is possible to recommend cross-genre, while Content-based filtering recommends similar contents.

However, despite its success and popularity, the traditional CF has some limitations. One of the critical limitations is the *item sparsity problem*. As the number of items in database increases, the density of each user's records with respect to these items will decrease [1]. If similar users have purchased different items, it is hard to know that those users have similar preferences. Another important limitation is the *cold start problem* (this problem is referred to as the *new item problem* in some

Jae-won Lee, Kwang-Hyun Nam, and Sang-goo Lee
School of Computer Science and Engineering
Seoul National University
Seoul 151-742, Republic of Korea
e-mail: {lyonking,nature1226,sglee}@europa.snu.ac.kr

R. Lee, N. Ishii (Eds.): Software Engineering, Artificial Intelligence, SCI 209, pp. 201–208.
springerlink.com   © Springer-Verlag Berlin Heidelberg 2009

papers). It is impossible to recommend or predict new or recently added items, until they are purchased by other users.

In this paper, we introduce a methodology for solving the item sparsity problem by mapping users and items to a set of concepts which are the nodes of domain ontology. By mapping them with concepts, the item sparsity can be reduced. That is, our method uses a semantic match with the domain ontology, while the traditional CF uses an exact match to find similar users. The rest of this paper is organized as follows. In Section 2, some previous works are reviewed, which are related to the similarity measures in collaborative filtering. In Section 3, our approach is explained. In Section 4, various experiments show that our approach is more precise than traditional CFs. Finally, we conclude this paper.

2 Related Work

Generally, CF is to predict the preferences of an active user from the preference data of other similar users. Therefore, it is the most critical to compute the similarity between the active user and other similar users exactly. In order to compute the similarity between users, the Pearson correlation coefficient [2] and cosine similarity [3] are widely used in traditional CF systems [4]. The Pearson correlation coefficient represents the angular difference between two users' preference vectors measured from the mean, whereas the cosine similarity measures the angular difference of two users' preference vectors that are measured from the origin.

The Pearson correlation coefficient and cosine similarity based CF use an exact match to compute the similarity between users, since these methods consider only explicit feedback on items such as users' scores, purchase logs, etc. [5]. That is, if the users purchased different items, they are considered as different users, which is the critical limitation of those methods. However, in our method items and users are semantically represented as a set of concepts. By representing their semantics with a set of concept, we can reduce the item sparsity problem and improve the precision of recommendation.

3 Semantics Based Collaborative Filtering

3.1 Overview

This section illustrates the process for recommending items with an example. In Figure 1 (a), user u purchased three items, such as i_1, i_3, and i_5, while user u' purchased two items such as i_2 and i_4. Although user u and u' have potentially similar preferences, they purchased different items. Therefore, they are regarded as different users (they have different preferences in traditional CF systems).

However, if items are presented as a set of concepts, the potential similarity between user u and u' can be computed because users are also represented as a set of concepts. In Figure 1 (b), although user u and u' have purchased different items, they have intersection concepts, such as c_1, c_2, c_5, and c_6. After computing the similarity between user u and u', the items, which user u' purchased but user u did not purchased, are recommended to user u.

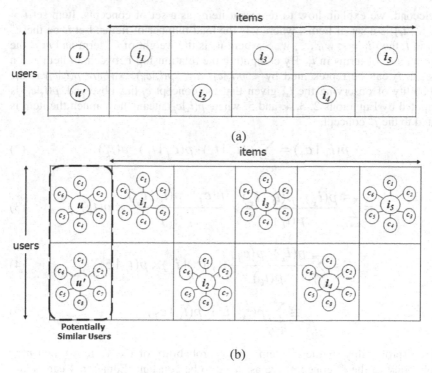

Fig. 1 (a) User-Item Matrix in Traditional CF and (b) User-Item Matrix in Semantic CF

3.2 Semantic Representation

This section explains how to represent items and users as a set of concepts. First, we explain how to model concepts used for semantic representation.[2]

Given ontology O, let $O = \{c_1, ..., c_n\}$ be a set of concepts. As each concept can be explained with Web pages that belong to the concept, c_i is the sum of Web page term vectors under c_i, which is computed as follows.

$$c_i = \frac{1}{|WP_i|} \sum_{wp_{ij} \in WP_i} wp_{ij} \qquad (1)$$

where WP_i is the set of Web pages that belong to the ith concept, and wp_{ij} is the jth Web page term vector in c_i. Each Web page wp_{ij} is represented as the weighted term vector $wp_{ij} = <w_{ij1}, ..., w_{ijm}>$ where m is the total number of terms in wp_{ij}. Each term weight w_{ijk} is computed using term frequency (TF) and inverse document frequency (IDF). When computing the term weights, we exploit a stop list to remove non-relevant terms from the description of c_i. In addition, Porter stemming [6] is used to reduce terms to their stems.

---

[2] Vectors are denoted by bold characters.

Second, we explain how to represent items as a set of concepts. Item set $I =$ $\{i_1, \ldots, i_v\}$ is a set of items, where v is the total number of items. Let i_x be the xth item in I, then $i_x = < w_1, \ldots, w_m >$ where w_k is the weight of kth term and m is the total number of terms in i_x. By computing the relationship between an item and a concept, i_x can be represented by $< p(i_x|c_1), \ldots, p(i_x|c_n)>$, where $p(i_x|c_j)$ is the probability of observing the i_x, given that the concept c_j has observed. $p(i_x|c_j)$ is computed by Equations 2, 3, 4, and 5, where $p(i_x|c_j)$ means how much the item is related to the jth concept.

$$p(i_x \mid c_j) = \sum_{t_k \in i_x} p(i_x \mid t_k) \cdot p(c_j \mid t_k) \cdot p(t_k) \tag{2}$$

$$= \sum_{t_k \in i_x} \frac{p(i_x) \cdot p(t_k \mid i_x)}{p(t_k)} \times \frac{p(c_j) \cdot p(t_k \mid c_j)}{p(t_k)} \times p(t_k) \tag{3}$$

$$= \sum_{t_k \in i_x} \frac{p(i_x) \cdot p(c_j)}{p(t_k)} \times p(t_k \mid i_x) \times p(t_k \mid c_j) \tag{4}$$

$$\cong \sum_{t_k \in i_x} p(t_k \mid i_x) \cdot p(t_k \mid c_j) \tag{5}$$

If $p(i_x)$ (probability of the xth item), $p(t_k)$ (probability of the kth term), and $p(c_j)$ (probability of the jth concept) are assumed to be constant, Equation 4 can be approximated to Equation 5. Moreover, $p(t_k \mid i_x)$ is computed by $w(i_x, t_k) / w(i_x)$, where numerator $w(i_x, t_k)$ denotes the term weight of t_k in i_x, the denominator $w(i_x)$ denotes the total weight sum of all terms in i_x. Similarly, $p(t_k|c_j)$ is computed by $w(c_j, t_k) / w(c_j)$, where numerator $w(c_j, t_k)$ denotes the term weight of t_k in c_j, and the denominator $w(c_j)$ denotes the total weight sum of all terms in c_j.

Third, we explain how to represent a user as a set of concepts. Let u be a user, $u = < p(u|c_j), \ldots, p(u|c_n) >$ where $p(u|c_j)$ is the probability of observing u, given that the concept c_j has observed. $p(u|c_j)$ is computed by Equations 6 and 7. The relationship of user u to the jth concept, $p(u|c_j)$, means how much the user preference is related to the concept c_j. $p(u|c_j)$ is computed as follows that is similar to $p(i_x|c_j)$.

$$p(u \mid c_j) = \sum_{i_x} p(u \mid i_x) \cdot p(c_j \mid i_x) \cdot p(i_x) \tag{6}$$

$$\cong \sum_{i_x} p(i_x \mid u) \cdot p(i_x \mid c_j) \tag{7}$$

If $p(i_x)$ is assumed to be constant, Equation 6 can be approximated to Equation 7. In Equation 7, $p(i_x|u)$ means how much the user prefers the item i_x, which is computed by $access(u, i_x)/access(u)$. $access(u, i_x)$ is the access count of u for i_x, and $access(u)$ is the total access counts of the user. The access count can be replaced with click-through count, rating score, or purchase counts depending on the properties of items.

3.3 Computing Semantic Similarity

Given the semantic representation for items and users, we need to explain how to find similar neighbors and how to generate a predicted value for the target items. Hereafter, u is an active user and u' is a similar user of u (we refer to u' as a target user). The semantic similarity between users is defined as follows.

$$p(u'|u) = \sum_{c_j} p(u'|c_j) \cdot p(u|c_j) \cdot p(c_j) \tag{8}$$

$$\cong \sum_{c_j} p(u'|c_j) \cdot p(u|c_j) \tag{9}$$

If $p(c_j)$ is assumed to be constant, Equation 8 is approximated to Equation 9. In Equation 9, $p(u'|c_j)$ and $p(u|c_j)$ are computed from Equation 7, respectively.

After computing the similarity between user u and u', the items with high predicted values are recommended to the active user u. The predicted value of item is defined as follows.

$$p(i_{u'x}|u) = \sum_{u'} p(i_{u'x}|u') \cdot p(u|u') \cdot p(u') \tag{10}$$

$$\cong \sum_{u'} p(i_{u'x}|u') \cdot p(u|u') \tag{11}$$

$$= \sum_{u'} p(i_{u'x}|u') \cdot p(u'|u) \tag{12}$$

In Equation 10, $i_{u'x}$ is the xth item which the user u' has accessed. Moreover, in Equation 11 and 12, if $p(u')$ is assumed to be constant, $p(u|u')$ can be replaced with $p(u'|u)$ which is proved through Equation 13 and 14.

$$p(u|u') = \sum_{c_j} p(u|c_j) \cdot p(u'|c_j) \cdot p(c_j) \tag{13}$$

$$= \sum_{c_j} p(u|c_j) \cdot p(u'|c_j) = p(u'|u) \tag{14}$$

4 Experiments

11,584 concepts from the Open Directory Project (ODP) whose domain is limited to music (/Top/Arts/ Music in ODP) are used as the nodes of domain ontology. The 50 users' click-through logs from Last.fm are crawled for experiments. In addition, 22,216 music meta-data are used as items.

For evaluation, we introduce some performance measures that are used hereafter. First, $precision_k$ is proposed. When we retrieve top-k items, we examine whether they are relevant to the user's preferences or not.

$$prcision_k = \frac{relevant \cap retrieved}{retrieved} \qquad (15)$$

$$= \frac{number\ of\ music\ the\ user\ clicked\ in\ top\ k}{k} \qquad (16)$$

In addition, we propose a new measure, *inverse rank precision$_k$* which is similar to *mean reciprocal rank*. *precision$_k$* means how much the system can reflect a user's whole preferences, while *inverse rank precision$_k$* means how much the system can reflect the user's main preferences. The closer *inverse rank precision$_k$* reaches 1.0, the more the search results are biased to the user's preferences.

$$inverse\ rank\ precision_k$$

$$= \frac{\sum_k \dfrac{1}{rank\ of\ music\ the\ user\ has\ listened}}{\sum_k \dfrac{1}{k}} \qquad (17)$$

4.1 Effects of Number of Similar Users

In this experiment, the effects of the number of similar users are explained when recommending items. The number of similar users is increased from 1 to 30. The chosen target users are the most similar to the active user. The proportion of training set is 50%. Precision (a) and inverse rank precision (b) at top-10 are measured. Figure 2 shows the effects of the number of users with *precision$_{10}$* and *inverse rank precision$_{10}$*.

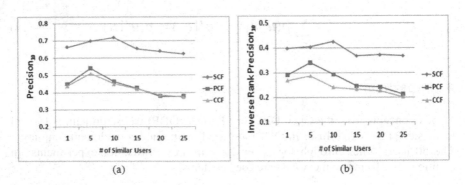

(a)   (b)

Fig. 2 (a) Precision$_{10}$ and (b) Inverse Rank Precision$_{10}$

In this experiment, *precision_{10}* and *inverse rank precision_{10}* are the highest when the numbers of similar users are 5 in PCF and CCF, but 10 in SCF. SCF indicates the Semantics based CF which is our approach, and PCF means the Pearson correlation coefficient based CF. Finally, CCF indicates the Cosine similarity based CF. The numbers of distinct items to recommend are sharply increased with the numbers of similar users in PCF and CCF, while the number of distinct items is slowly increased in SCF.

Therefore, the numbers of similar users of PCF and CCF are smaller than that of SCF. The result also shows that, in all cases, SCF provides more accurate performance than PCF and CCF because by mapping users and items to a set of concepts, SCF can reduce the sparsity of items. Particularly, though there is a little difference in *precision_{10}* of PCF and CCF, *inverse rank precision_{10}* of CCF is much lower than that of PCF in Figure 2 (b). Since the Pearson correlation coefficient considers the differences in users' preference scales, PCF is a little bit more accurate than CCF.

4.2 Effects of User Type

In this experiment, distinct 16 users' profiles among 50 profiles are chosen and categorized into four types. The criteria for categorizing users are the number of music a user has clicked and the total click-through count of the user as shown in Figure 3.

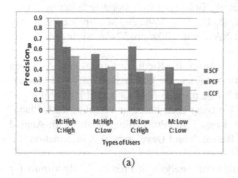
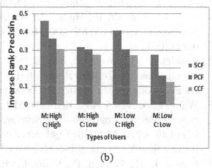

(a)   (b)

Fig. 3 (a) Precision_{10} and (b) Inverse Rank Precision_{10}

The numbers of similar users are 5 in TCF and CCF, while the number is 10 in SCF. In Figure 3, if the number of music the user has clicked (denoted by 'M') and the total click-through count of the user (denoted by 'C') are high, *precision_{10}* and *inverse rank precision_{10}* are the highest. If both criteria are low, *preicision_{10}* and *inverse rank precision_{10}* are also the lowest. In SCF, the effect of total click-through count dominates the performance rather than the effect of number of music. That is, the total click-through count reflects the user's preferences accurately compared to the number of music the user has clicked. However, the number of music the user has clicked is a more critical factor that determines the overall performance in PCF and CCF.

5 Conclusion

By mapping users and items to a domain ontology, we improved the precision of recommendation and solved some problems of the traditional CF. We explained how to map items and users to a set of concepts with a probabilistic model. The concepts that are mapped to a user's profile are learned from the user's history. Two main contributions are (1) mapping items and users to a set of concepts that are the nodes of ontology, and (2) demonstrating that our method is more precise than traditional CF methodologies. Our future work includes the improvement of performance by proposing hybrid CF, which consists of the Semantic Collaborative Filtering and the Semantic Content based Filtering.

Acknowledgement. This research was supported by the Ministry of Knowledge Economy, Korea, under the Information Technology Research Center support program supervised by the Institute of Information Technology Advancement. (grant number IITA-2008-C1090-0801-0031)

This research was supported by a grant (07High Tech A01) from High tech Urban Development Program funded by Ministry of Land, Transportation and Maritime Affairs of Korean government.

References

1. Mobasher, B., Jin, X., Zhou, Y.: Semantically Enhanced Collaborative Filtering on the Web. In: Berendt, B., Hotho, A., Mladenič, D., van Someren, M., Spiliopoulou, M., Stumme, G. (eds.) EWMF 2003. LNCS (LNAI), vol. 3209, pp. 57–76. Springer, Heidelberg (2004)
2. Resnick, P., Lacovou, N., Suchak, M., Bergstrom, P., Riedl, J.: GroupLens: An Open Architecture for Collaborative Filtering of Netnews. In: Proceedings of ACM Conference on Computer Supported Cooperative Work, pp. 175–186. Chapel Hill (1994)
3. Wang, J., de Vries, A.P., Reinders, M.J.: Unifying User-based and Item-based Collaborative Filtering Approaches by Similarity Fusion. In: Proceedings of the 29th Annual international ACM SIGIR Conference on Research and Development in information Retrieval, New York, pp. 501–508 (2006)
4. Breese, J., Heckerman, D., Kadie, C.: Empirical Analysis of Predictive Algorithms for Collaborative Filtering. In: Uncertainty in Artificial Intelligence. Proceedings of the Fourteenth Conference, pp. 43–52. Morgan Kaufman, San Francisco (1998)
5. Lee, T.Q., Park, Y., Park, Y.T.: A Similarity Measure for Collaborative Filtering with Implicit Feedback. Advanced Intelligent Computing Theories and Applications with Aspects of Artificial Intelligence, 2007, 385–397 (2007)
6. Porter, M.F.: An algorithm for suffix stripping. In: Sparck Jones, K., Willett, P. (eds.) Readings in information Retrieval. Morgan Kaufmann Multimedia Information And Systems Series, pp. 313–316. Morgan Kaufmann Publishers, San Francisco (1997)

Evaluation and Certifications for Component Packages Software

Haeng-Kon Kim

Summary. There is a growing interest in tools and packages that support the capture, representation and gradual refinement of models of an organizations software processes, especially in the light of schemes. It is confined to definition, execution, and measurement of the process in the process life cycle while its application package scope is the overall business process within the company. In order to comprehensively manage and improve business process, collecting accurate data on software packages is needed.

Package software should have the feature that purchasers can discriminate a product suitable for them among a number of software, which belong to the similar kind of product. Purchasers' ability to choose a package software depends on that they can judge whether a package software has the relevant standard conforming through objective quality test process and method or not. For building this system, there are the standards that can be applicable to package software, such as <ISO/IEC 14598-5 : Quality Evaluation Process for Evaluator>and <ISO/IEC 12119 : Information Technology - software package - Quality Requirements & Test>. In this paper, we built the system that purchasers can effectively select a package software suitable for their needs, building quality test and certification process for package software and developing Test Metric and application method.

Keywords: component package, quality evaluation, package software.

1 Introduction

The development and maintenance of a project plan for a software packages is a complex activity. Current project planning tools for the software packages provide little more than task scheduling functionality. Project planning

Haeng-Kon Kim
Department of Computer Information Communication Engineering,
Catholic University of Daegu, Korea
hangkon@cu.ac.kr

R. Lee, N. Ishii (Eds.): Software Engineering, Artificial Intelligence, SCI 209, pp. 209–220.
springerlink.com   © Springer-Verlag Berlin Heidelberg 2009

tools for the packages that contain explicit representations of knowledge of relevant standards, methods and best practice would provide significantly greater support for creating and refining a software development plan. Due to the rapid spread of personal computers, a variety of package software for personal or office use have been developed, and consequently the liberty of choice has been broadened. Package software should have the feature that purchasers can discriminate a product suitable for them among a number of software, which belong to the similar kind of product. If we want to make a right choice for package software, we should consider whether a package software satisfies the established standard or not through objective quality test process and method.

For building this system, there are the standards that can be applicable to package software, such as <ISO/IEC 14598-5 : Quality Evaluation Process for Evaluator>and <ISO/IEC 12119 : Information Technology - software package - Quality Requirements & Test>. In case of ISO/IEC 12119, those can use it, such as software developers, organizations for authentication that intend to establish third-party authentication, organizations for approving authentication and test centers, and software purchasers.

In this paper, we developed the method that can contribute to quality improvement of package software by building the quality test process for package software based on this standard and developing test metric and application method. This study introduces the present research state related to quality in Chapter 2, and builds the test process for package software from the purchasers' viewpoint in Chapter 3. It introduces quality model for testing package software in Chapter 4, and describes the metric that was developed based on quality model in Chapter 5, and finally describes the conclusion and further studies.

2 Related Works

2.1 Software Process Modeling

A software process model is an abstract representation of software development activities. Software life-cycle models such as the waterfall and V-models are examples of simple software process models that focus on the high-level activities and products involved in software development. In practice, they are too abstract to be of much practical help other than for illustrating some of the dependencies between development activities and products. Organizations will often have much more detailed software development processes documented in a quality manual or specified by a methodology. However, these representations are mainly implicit and informal and can be difficult to communicate and analyze.

Software process modeling and the technology that supports it are relatively recent developments, but ones with important potential benefits. It supports:

- *understanding* and *communication* of development processes by helping resolve ambiguities and allowing an organization to communicate and share best practice for re-use on future projects;
- *assessment* of processes by allowing comparative analysis of different projects and by assessing the maturity and capabilities of an organizations development processes;
- *process improvement* by helping to identify problem areas and estimating the impacts of potential changes;
- *project planning* by facilitating the development of project specific process plans to meet project specific characteristics while satisfying development standards and methodologies;
- *process enactment* by automating portions of the development process, supporting co-operative work of the development team, and helping to collect metrics about the development process.

There are several schemes and programmers that are either explicitly addressing the need for software process modeling or at least generating interest in the topic. The UK TickIT scheme [3], a software sector specific version of ISO9001, requires people to document their software development processes. SPICE (Software Process Improvement and Capability dEtermination) [4] is an ISO backed project which aims to develop an international standard for software process capability determination. The project is drawing upon ideas from the US Software Engineering Institutes Capability Maturity Model (CMM) [5] and the UK TickIT scheme. In the US, ARPA (Advanced Research Projects Agency) is sponsoring the STARS (Software Technology for Adaptable, Reliable Systems) [6] programme. STARS is focused on improving the way software is developed within the US DoD. One of its aims is to accelerate the transition to a process-driven, technology supported software development paradigm.

2.2 6-Sigma

6-Sigma thus emerged as the best management innovation strategy in the history of quality management. The business sector has seen a continual improvement of quality management systems, which evolved through five separate eras: Quality Control (QC), Statistical Quality Control (SQC), Quality Assurance (QA), and Total Quality Management (TQM). 6-Sigma brought about radical reduction of costs, competitiveness enhancement, and changes in organizational culture. 6-Sigma also integrated and links with other management techniques, thereby continuously evolving itself and expanding to other industrial areas for implementation.

The key elements of the third generation 6-Sigma are as follows;

① Allow participation of as many people engaged in a company's value chain.
② Spread 6-Sigma in broad scale by the use of new technologies such as the Internet.
③ Let the strength of 6-Sigma come from all employees, not a small number of experts.

2.3   Current Trends in Domestic and Foreign

A. Foreign Trend
Foreign advanced countries in software are continuously trying to establish the standard for quality evaluation. They are on the way to standardize ISO/IEC 9126 as the standard on quality evaluation features and ISO/IEC 14598 as the standard on quality evaluation process. However, it is the actual circumstances that it is very rare they build the specific quality evaluation method and then actually apply it, based on the general contents on standard. And there is a case that they build the practical evaluation system about application, a part of quality features, and then utilize it.

B. Domestic trend
It can be said that now the domestic trend on quality evaluation test technology has its weak basis on the whole. The related standard for quality evaluation has not been established yet, and the authentication for software's quality system relies on foreign countries, and thereby we can see the basic study in domestic is very weak. Even though domestic software industry regards technology for quality improvement and development of product evaluation technology as the urgent task, it has much difficulty in pushing technology development in itself.

3   Building Certification Process for Package Software

3.1   The Outline of the Process

The outline of the test/certification process is like figure 1. We constructed it 'certification request and acceptance', 'certification', 'certification announcement and delivery' and 'activity after certification'. The test during the certification make progress through the self test department or an external test department.

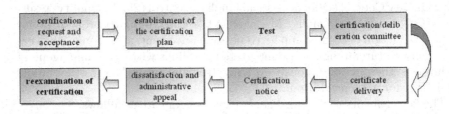

Fig. 1 The Outline of the Test/Certification Process

3.2   The Detailed Activity of the Process

(1) Certification Request and Acceptance

The certification applicant presents software quality certification request form, product descriptor, user's manual and software as in figure 2.

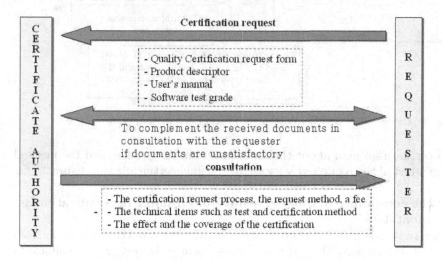

Fig. 2 Certification request and acceptance

The certificate authority presents the following items to the certification requester.

- The certification request process, the request method, a fee
- The technical items such as test and certification method
- The effect and the coverage of the certification

The certificate authority can analyze certification request documents and complement the received documents in consultation with the requester if they are unsatisfactory

(2) test asking and establishment of the certification plan

The certificate authority confer the following items with the requester to implement the certification.

- The days program and the method of the certification
- Required items when the certification is implemented
- Etc.,

When the test is requested with a certification, the certification center requests the software test to the test department with 'software test request form', 'product descriptor', 'user's manual' and software and is informed the written agreement by the test department. And the certification center make out

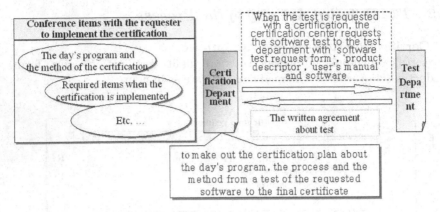

Fig. 3 Test Request and Establishment of the Certification plan

the certification plan about the day's program, the process and the method from a test of the requested software to the final certificate as in figure 3.

(3) the acceptance of the test results and framing of certification assessment data

When the test is ended, the certification team is delievered the test results from the test department. The test results must include the test grade to inform to the test requester. The certification team starts assessment task analyzing the test results and related data, and make out the certification/assessment data dependant on the criteria of certificate authority.

(4) certification/deliberation committee and assessment report

When the certification/assessment data were completed, the certificate authority hold a certification/deliberation committee and present agenda. The Certificate authority make out the assessment report based on the result of certification/deliberation committee and certification/assessment data and inform it to the applicant with the test grade after assessment.

When necessary, the certificate authority specify inadequate items, a additional assessment, a test coverage and approvement measures. If requester proof to take steps for improvement meeting all requirements in a given period, they redo the required assessment.

(5) certificate delivery and notice

After certification/assessment, if the software applied for certification is coincided with the certification condition and certification/deliberation committee decide resolve the certification, the certificate authority deliver a certificate and announce the certification.

The announced contents are as follows.

- The Category of Certification
- A firm name of those who is certified

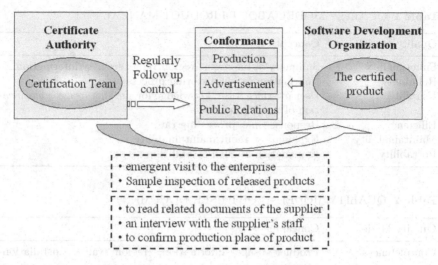

Fig. 4 Follow up control

- Software name and version
- Certificate number
- Certificate number

(6) dissatisfaction and administrative appeal
When the client's dissatisfaction or demur is appealed, the certificate authority receive by a written application. If the received contents is slight, the certificate authority take immediate action by document. If they are important, the certification/deliberation committee treat it as in figure 4.

(7) follow up control
The certificate authority regularly control product sale, advertisement, production. If software affect to conformance of the certified software, the supplier must inform the certificate authority.

4 Quality Model

In order to apply ISO/IEC 12119 to package software test, the quality model, which each item consisting of package software is to be applied to, should be organized.

4.1 Quality Model on Product Manual

Quality model on product manual among the constitutional elements of package software includes the items such as function, reliability, application, effectiveness, maintenance and graft, and those can be summarized as shown in Table 1.

Table 1 QUALITY MODEL ABOUT PRODUCT MANUAL

Quality Model	Concept
Functionality	Summary of functions, region value, security information
Reliability	Information for data storing process
Usability	User interface form, knowledge for product usage, identification of usage condition
Efficiency	Response time, processing rate
Maintainability	Explanation about maintainability
Portability	Explanation about Portability

Table 2 QUALITY MODEL ABOUT USER DOCUMENT

Quality Model	Concept
Completeness	Product usage information, region value, installation-maintenance manual
Correctness	Correctness of document information, clearness of expression
Consistency	Integrity between documents, terms consistency
Understandability	User group have to understand
Easy summary	Easy summary about user documenet

4.2 Quality Model on User Document

Quality model on user document among the constitutional elements of package software includes the items such as perfection, exactness, consistency, understanding and easy summary and those can be summarized as shown in Table 2.

Table 3 QUALITY MODEL ABOUT PROGRAM AND DATA

Quality Model	Concept
Functionality	- Can Install according to the manual - similiar to all explanation in other document - not conflict with other documents - must be executed as specification
Reliability	- always controllable - data is not destructed
Usability	- understandability about all information of program - adequacy of error message information
Efficiency	- the explanation about efficiency is suitable
Maintainability	- the explanation about maintainability is suitable
Portability	- the explanation about portability is suitable

4.3 Quality Model on Program and Data

Quality model on program and data among the constitutional elements of package softwares includes the items such as function, reliability, application, effectiveness, maintenance and graft, and those can be summarized as shown in Table 3.

5 Development of Package Software Evaluation Metrics

Evaluation Metric for package softwares has the basis of ISO/IEC12119, and this study abstracted the Metric items that are applicable to package softwares from ISO/IEC 9126-2, 3, and modified and supplemented them. The details of developed Metric items are as shown in Table 4.

5.1 Metric Index Table

This study built the Metric Index Table by product element consisting of package software, as shown in Table 5. The Metric Index Table on general requirements for package software is shown in the Table.
Metric Table was developed, based on ISO/IEC 12119, and it was modified as suitable one for package softwares test by introducing some relevant items from ISO/IEC 9126-2, 3.

Table 4 QUALITY MODEL WITH METRIC

Characteristics	Metric index	Type	Reference
	1.1 identification of product manual	Y/N	ISO/IEC 12119
	1.2 identification of product	Y/N	ISO/IEC 12119
Identification and order	1.3 Specification of supplier	Y/N	ISO/IEC 12119
	1.4 Specification of work	Y/N	ISO/IEC 12119
	1.4 Document for adequacy requirements	Y/N	ISO/IEC 12119

Table 5 A SAMPLE OF METRICS TABLE ABOUT GENERAL REQUIRE-
MENTS

Quality characteristics	Identification & order	
Specification of metric	Detail item	Measurement result value
1.1 identification of product manual computation : A value range : 0, 1 A	Is a unique document ID in product manual?	
	Example : name for product manual (function manual, product information, product pamplet, etc.	
problems		
1.2 identification of product computation : A value range : 0, 1 A	Is a unique ID in software product?	
	(Example) name, version, date, variant (Example) Variant : Enterprise version, Professional version, etc.	
problems		

5.2   Construction of Metric Table

An example of Metric that is developed for the purpose of testing package softwares by product element is as shown in Table 6. The example of Metric on general requirements for package softwares is shown in the Table.

5.3   Decision of the Evaluation Marks Level and Judgement Standard on Metric value

If the result value intends to have the meaning, it needs to decide the evaluation marks level on Metric value.

First, we define the evaluation marks level by deciding the number of range that Metric value has. The following example shows the case that defines 4 evaluation marks levels.

Table 6 REMARK OF METRICS

Type of Metrics	The number of metrics	Remark
General requirements	10	Metrics about Identification and order
Product manual	20	Metrics about functionality, reliability, usability, efficiency, maintainability, portability
User document	12	Metrics about completeness, correctness, consistency, understandability, easy summary
Program & data	61	Metrics about functionality, reliability, usability, efficiency, maintainability, portability

A : excellent : satisfy all requirements
B : good : satisfy almost requirements
C : fair : not satisfy a part of requirements
D : poor : not satisfy requirements

We can decide the range corresponding to evaluation marks level on each Metric value as follows.

Measurement value $0<=X<=1$
$X<0.6$: rating level D
$0.6<=X<0.7$: rating level C
$0.7<=X<0.8$: rating level B
$0.8<=X$: rating level A

Since the range of Metric measurement value is not always fixed, we decide it by considering the range of measurement value on each Metric. In this way, we can score evaluation marks according to evaluation marks level on each Metric, and if it acquires a certain level of evaluation marks, we get the final result by deciding the criterion to pass or fail. For example, supposing that they decide to purchase if the number of Metric, of which the evaluation marks level is B or above, is 95% or more, and if the test is applied to the several software as objects, we can decide to purchase the software that acquired the best result.

6   Conclusion

This study built quality test process for package software and developed Metric for testing package software by attempting to graft product evaluation process for evaluator in ISO/IEC 14598-5 into the standard of quality test for package software in ISO/IEC 12119, considering the features of package software.

If we firmly build evaluation system for package software with basis of the process for evaluator in ISO/IEC 14598-5, it is considered that we can build the effective evaluation basis for package software types that are made by many development organizations.

Regarding the study works after this, it needs to specify measurement methods on measured items of test Metric for package software, and push to develop effective quality test through tools.

Acknowledgements. This research was supported by the MKE(Ministry of Knowledge Economy), Korea, under the ITRC(Information Technology Research Center) support Program supervised by the IITA(Institute of Information Technology Advancement) (C1090-0902-0032).

References

1. ISO/IEC 9126, Information Technology-Software Quality Characteristics and metrics-Part 1, 2, 3
2. ISO/IEC 14598, Information Technology - Software product evaluation - Part 1, 2, 3, 4, 5, 6
3. Moller, K.H., Paulish, D.J.: Software Metrics. Chapmen & Hall, IEEE Press (1993)
4. Wallmuller, E.: Software Quality Assurance A practical approach. Prentice-Hall, Englewood Cliffs (1994)
5. Harry, M.J., Schroeder, R.: Six Sigma. Bantam Dell Pub. Group (1999)
6. Park, J.S., Kim, O.S., Kim, J.C., Jeon, B.G.: G100 Stories of Six Sigma for GB, BB and MBB, Korean Standard Association (2004)
7. Yang, H.-S., Lee, H.-Y.: Design and Implement of Quality Evaluation Toolkit in Design Phase. KISS Paper(C) 3(3) (1997)
8. Yang, H.-S.: Quality Assurance and Evaluation of Hanjin Shipping New Information System. Hanjin Shipping co. (1998)
9. Yang, H.-S.: Development of Software Product Evaluation Supporting Tool, ETRI Computer Software Technology Institute (1999)
10. Yang, H.-S.: Study on Quality Test and Measurement Criteria, ETRI, Final Report (2005)
11. Yang, H.-S.: Construction of Quality Test and Certification System for Medical soft-ware, Korea Food & Drug Administration, Final Report (2005)
12. Wigg, K.M.: Integrating Intellectual Capital and Knowledge Management. Long Range Planning 30(3), 399–405
13. Koh, J.W.: Comparison and Evaluation of Knowledge Storage Model Performance of Knowledge Management Organization, MS Thesis, KAIST (2004)
14. Shin, H.J.: Study of Development of E-AgentComponent for e-Business Application System Support, PhD Dissertation, Daegu Catholic University Graduate School (2003)
15. Song, D.H.: Study of E-AgentBased Teaching-Learning Cases and Applications, MS Thesis, Kyeonggi University Graduate School of Education (2001)
16. Wooldridge, M.: Agent-based Software Engineering. IEEE Proceedings on Software Engineering 144(1), 26–38 (1997)

Categorical Representation of Decision-Making Process Guided by Performance in Enterprise Integration Systems

Olga Ormandjieva, Victoria Mikhnovsky, and Stan Klasa

Abstract. The research work presented in this paper is motivated by the need to build performance measurement and decision making process into the enterprise models, a need rooted in the current industrial trend toward developing complex integrated enterprises. In order to accomplish Enterprise Integration (EI) compliance with the imposed performance requirements we formalize the EI modeling and performance control in a single formal framework based on representational theory of measurement and category theory. Category theory is expressive enough to capture qualitative and quantitative knowledge about heterogeneous EI requirements, their interrelations and decision-making mechanism in one formal representation, where structure and reasoning are inextricably bound together. Thus, category theory provides a computational mechanism whereby such knowledge can be applied to EI data and information structures to arrive at conclusions which are valid.

Keywords: Performance modeling, enterprise integration, category theory, representational theory of measurement, decision-making.

1 Introduction

Enterprise Integration (EI) occurs when there is a need for improving interactions among people, systems, departments, services, and companies (in terms of material flows, information flows, or control flows). Enterprises typically comprise hundreds, if not thousands, of applications (comprising of packaged solutions, custom-built, or legacy systems), some of them being remotely located. The research work presented in this paper is motivated by the need to build performance measurement and decision making process into the enterprise models, a need rooted in the current industrial trend toward developing complex integrated enterprises. The importance of enterprise integration compliance with the imposed performance requirements necessitates continuous control of the performance indicators in real time in increasingly complex large-scale EI systems.

Olga Ormandjieva, Victoria Mikhnovsky, and Stan Klasa
Department of Computer Science & Software Engineering, Concordia
University, Montreal, Canada
e-mail: `ormandj@cse.concordia.ca`, `tori.mikhnovsky@hotmail.com`,
`klasa@cse.concordia.ca`

R. Lee, N. Ishii (Eds.): Software Engineering, Artificial Intelligence, SCI 209, pp. 221–232.
springerlink.com   © Springer-Verlag Berlin Heidelberg 2009

Research goal. The goal of this work is to formalize the EI modeling and performance control in a single formal framework based on representational theory of measurement and category theory.

Our approach. Enterprise architecture modeling will need a mechanism to acquire and represent the structure of enterprise components and their interrelations, and map them onto performance measurement structure which is a hierarchy of measureable indicators in line with enterprise decision-making strategy and objectives. Category theory is a branch of mathematics which is supremely capable of addressing "structure"; structure emerges from interactions between elements as captured by arrows, and not extensionally as in set theory [2, 3, 4]. Thus, it is suitable for formalizing system-level interactions between enterprise components, and controlling system behavior's conformance to performance requirements for decision-making purposes. Through the application of category theory we seek to avoid the undesirable results of informal and sometimes arbitrary assignment of numbers as proposed by many measurements [5,6,7]. Using the categorical theoretical framework we analyze the mathematical mappings from the empirical world into numbers and the properties of those mappings. The importance of the mappings are at least threefold: (1) understanding empirical properties of mapping, (2) detecting undesirable properties of mapping, and (3) correctly using the resulting numbers in statistical and related performance analysis required for enterprise integration decision-making.

A major application of measurement theory is to facilitate decision-making. Putting measurement on a firm foundation plays a very important role in, for instance, risk management based on judgments on EI performance.

Contributions. The results described in this paper contribute to the research in the EI and performance measurement community through exploiting the applicability of category theory as a unifying formal language that allows using the same constructs for modeling heterogeneous objects and different types of relations between them. Our novel categorical framework is expressive enough to capture knowledge about heterogeneous requirements, their interrelations and decision-making mechanism in one formal representation, where structure and reasoning are inextricably bound together.

The rest of the paper is organized as follows. Background and motivation for this research is given in Section 2. Section 3 presents Enterprise Integration (EI) and performance metamodel with decision-making. Categorical representation of EI is detailed in Section 4. Section 5 presents the conclusions and further research directions.

2 Background and Motivation

The background to this research extends back to the mid-1980s when the need for better-integrated performance measurement systems was identified [8,9,10,11,12]. Since then there have been numerous publications emphasizing the need for relevant, integrated, balanced, strategic and improvement-oriented performance measurement systems. None of the existing studies [13,14,15,16,17,18,19,20,21] consider performance measurement from an enterprise integration perspective and in order to identify the most appropriate performance measurement system for enterprise integration, a review of performance measurement systems in supply chains, extended enterprises and virtual enterprises was conducted [22,23,24]. The majority of the EI measurements proposed so far have been adopted from the traditional measurement procedures of single enterprises corresponding to supply chains (i.e. plan, source, make, deliver, etc.).

Our background study into performance measurement in EI concluded that the goal of EI is to build a knowledge-based organization, which uses the distributed capabilities, competencies and intellectual strengths of its members to gain competitive advantage to maximize the performance of the overall intergraded enterprise. Nevertheless, despite the large expenditures on EI, few companies are able to assess exactly the current level of performance of their business. Performance measurement at an individual enterprise level plays only a minor role, and at the integration level it is virtually nonexistent. Furthermore, none of the current strategic models and frameworks for performance measurement, such as balanced scorecard, performance prism, IPMS, smart pyramid etc., consider performance measurement and management from enterprise integration perspective. Other works on performance measurement in supply chains, extended enterprises and virtual enterprises specify a range of performance measures, which should be used in managing supply chains and virtual organizations but fail to integrate them within a strategic performance measurement framework.

Inter-enterprise coordinating (or partnership) measures are essential to ensure that various partners within an extended enterprise coordinate effectively and efficiently to ensure that the performance of the integrated enterprise is maximized. Unfortunately, none of the current strategic models and frameworks for performance measurement explicitly considers the need for inter-enterprise coordinating measures. Moreover, within these frameworks we did not find a structured method that establishes solid vinculums among the different performance measurement elements to achieve coherent traceability among elements. Thus, existing frameworks fail to deliver a robust structure throughout all levels of performance indicators: project and product, functional, individual enterprise, and integrated enterprise.

In practice, there are very few frameworks related to performance management based on the measurement theory. Moreover, the frameworks that address measurement in a formal way leave out analyzing the mathematical mappings from the empirical system into a numerical system and properties of those mappings [1].

Category theory is a relatively young branch of mathematics designed to describe various structural concepts from different mathematical fields in a uniform way [2,3,4]. Category theory offers a number of concepts and theorems about those concepts, which form an abstraction and unification of many concrete concepts in diverse branches of mathematics. Category theory provides a language with a convenient symbolism that allows for the visualization of quite complex facts by means of diagrams. Enterprise engineering is very fragmented, with many different sub-disciplines having many different schools within them. Hence, the kind of conceptual unification that category theory can provide is badly needed. As pointed out by Goguen [25], category theory can provide help in dealing with abstraction and representation independence. In enterprise integration, more abstract viewpoints are often more useful, because of the need to achieve independence from the overwhelmingly complex details of how things are represented or implemented.

Despite the popularity of category theory in some fields of computing science, no applications in the field of modeling EI or business process can be found in the literature with the exception of authors earlier work presented in [26, 27].

3 Enterprise Integration and Performance Metamodel with Decision-Making

The background to this research extends back to the mid-1980s when the need for better-integrated performance According to the ISO/IEC 25030:2007 International Standard [28], enterprise system is a hierarchy of systems including information systems, human business processes and communication among them. Figure 1 visualizes the generic enterprise system hierarchical model that forms a primary taxonomy for analyzing communication relationships between EI entities.

We abstract EI to a collection of communicating business processes. In our EI model we consider only one way of communication between business processes: events. The information system services which are triggered by the events included in a single business process may come from more than one computer system; the communication between computer systems providing the information system services is irrelevant to this model and, thus, is hidden.

Because non-functional requirements like performance tend to have a wide-ranging impact on enterprise architecture, existing enterprise modeling methods are incapable of integrating them into the enterprise engineering process.

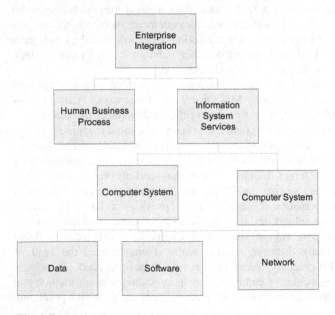

Fig. 1 Enterprise Integration Hierarchy

Performance requirements impose restrictions on the integration by specifying external constraints on the business processes and therefore need to be considered as an integral part of the EI metamodel.

The performance model proposed in this paper serves as a framework to ensure that all aspects of quality are considered from the internal and external points of view. It covers all performance aspects of interest for most enterprise systems and as such can be used as a checklist for ensuring complete coverage of performance in the integration process.

In our approach performance is modeled as a hierarchical information structure (see Figure 2). In each of the decompositions the offspring (sub) characteristics can contribute partially or fully towards satisfying the parent.

The lowest level corresponds to quantifiable performance sub-characteristics of EI entities (here, business process, project, information system service, enterprise integration) quantified by applying a measurement method - a logical sequence of operations applied directly on the source, that is, an entity. The result of applying a measurement method is called a base measure; base measures are then combined by a measurement function to obtain derived measures required for characterizing the parent (that is, indicator) in accordance with the associated rules for interpretation of measurement data [29] (see Figure 2). An indicator provides an estimate or evaluation of specified attributes derived from the analysis of the measurement data (values) with respect to defined decision criteria, which serves as basis for decision-making by the measurement users. For example, acceptable range of software reliability values is [75%, 100%]; and values below 75% would require more testing of the product until an acceptable level is reached.

To sum up, performance indicators are quantified by measurement procedures providing measurement methods, functions, and a meaningful analysis algorithm for combining measurement data and decision making criteria.

EI performance requirements can be modeled and measured along many dimensions, each corresponding to a certain indicator and its decomposition into the offspring measures. A measurement planner defines measurement indicators according to decision-making information needs. In our research [26,27] we used

Fig. 2 Performance Measurement Hierarchy

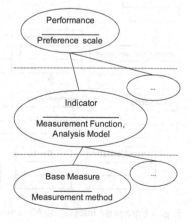

the goal-driven approach to develop a hierarchical Enterprise Performance Measurement Model (EPMM) which ensures that all aspects of performance are considered from the internal and external points of view. EPMM covers all aspects of interest for most enterprise systems and as such can be used as a checklist for ensuring complete coverage of performance in the integration process. The model allows the performance to be decomposed into a hierarchy of measurable indicators in line with enterprise decision-making strategy and objectives. The performance of EI is decomposed into eight qualitative performance factors: flexibility, coordination, dependability, quality, timing requirements, cost, asset utilization, and capacity. The factors are repetitively refined further into quantitative indicators. For example, dependability can be decomposed into product and operational dependability; two important dimensions of operational dependability are: availability – the ability of a process to deliver products and services when requested, and reliability – the ability of the process to deliver high-quality products and services under the pressure of environmental uncertainties. The EPPM details are provided in [26,27] and omitted from this paper due to the lack of space.

The knowledge about EI hierarchy (business processes, events, information system services), performance hierarchy of measurable indicators in line with enterprise decision-making strategy and decision-making rules, and the relationships among them are captured in the metamodel depicted in Figure 3.

One of the main issues in EI performance measurement and quality measurement in general is the validity of the measurement procedure. The theoretical validity of the measurement methods is guaranteed by the representational theory of measurement [1,30]. A major application of the representational theory of measurement is to ensure valid and reliable decision-making process.

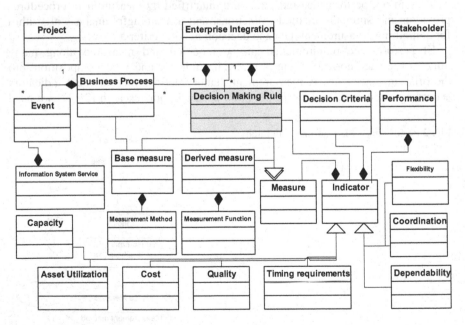

Fig. 3 Enterprise Integration Metamodel

With the purpose of accomplishing EI compliance with the imposed performance requirements, we formalize the EI modeling and performance decision making in a uniform formal framework based on representational theory of measurement and category theory.

4 Generic Categorical Representation

This section gives a brief summary of the categorical point of view on the EI structures explained above and their relations. The detailed description and analysis of the categorical representation of EI, performance measurement and decision-making will be available in a forthcoming extended version of this work.

4.1 Some Basic Category Theory Notions

Categorical notations consist of diagrams with arrows. A category consists of a collection of objects and a collection of arrows (called morphisms). Each arrow f: $A \rightarrow B$ represents a function. Representation of a category can be formalized using the notion of the diagram. Although category theory is a relatively new domain of mathematics, introduced and formulated in 1945 [31], categories are frequently found in this field (sets, vector spaces, groups, and topological spaces all naturally give rise to categories). The use of categories in Enterprise Integration can enable the recognition of certain regularities in distinguishing performance in a variety of business processes, their interactions can be captured and composed, equivalent interactions can be differentiated, patterns of interacting business processes can be identified and some invariants in their action extracted, and a complex object can be decomposed into its basic components. Moreover, automation may be achieved in category theory, for example, the composition of two business processes can be derived automatically and the category of business processes follows some properties, such as cooperation. Category theory for EI can adopt a correct by construction approach by which components are specified, proved, and composed in the way of preserving their properties.

A category consists of objects: A, B, C, etc., and arrows (morphisms) $f : A \rightarrow B$ where for each arrow f, there are given objects: $dom(f)$, $cod(f)$ called domain as well as codomain of f, and indicated as $A = dom(f)$ and $B = cod(f)$ respectively. Central to the category theory is the notion of composition : given arrows $f : A \rightarrow B$ and $g : B \rightarrow C$ with $cod(f) = dom(g)$, there is an arrow: $g \circ f : A \rightarrow C$ called a *composite* of f and g. For each object A, there is a given arrow: $id_A : A \rightarrow A$ called *identity arrow* of A. The category must satisfy the following laws: i) Associativity: $h \circ (g \circ f) = (h \circ g) \circ f$ for all $f : A \rightarrow B$, $g : B \rightarrow C$, $h : C \rightarrow D$; ii) Unit: $f \circ id_A = f = id_B \circ f$ for all $f : A \rightarrow B$.

A functor $F: C \rightarrow D$ between categories C and D is a structure-preserving mapping of objects to objects along with arrows to arrows in the way of: i) $F(f: A \rightarrow B) = F(f) : F(A) \rightarrow F(B)$; ii) $F(g \circ f) = F(g) \circ F(f)$; iii) $F(id_A) = id_{F(A)}$

In this section we follow the categorical modeling of general relations proposed in [32].

4.2 Representational Theory of Measurement v.s. Category Theory

Making judgments based on performance indicators should satisfy certain conditions (axioms) about an individual's judgments, such as preferences, which make the measurement possible. To perform a measurement, we start with an observed or empirical relational structure E, representing entities to be judged (like, for instance, business processes in an enterprise integration), and we seek a mapping (here, performance measurement) to a numerical relational structure N, which "preserves" all the relations and operations in E. A mapping from one relational structure to another that preserves all relations is called a homomorphism (see Figure 4).

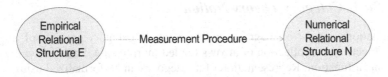

Fig. 4 Representing Measurement

Let E be a nonempty set and $R \subseteq E \times E$ be a binary relation on E denoted by (E, R). As usual, we express a relationship $(x,y) \in R$ between two elements $x, y \in E$ by xRy. In category theory, such relation xRy is visualized by an arrow $x \rightarrow y$; such an arrow will be interpreted as a morphism between objects x and y.

The empirical relational structure is modeled as a partially ordered set ("poset") $(E, \leq)$ where $\leq$ is reflexive, transitive and antisymmetric. We can associate with this poset the category E (for instance, *Business Process*, see Figure 5). The objects of E are the elements of the empirical structure (i.e., in the category *Business Process* the objects are business processes) and for BP1, BP2 in *Business Process* there is a morphism $BP1 \rightarrow BP2$ if and only if $BP1$ and $BP2$ are in relation $\leq$ (such as preference), i.e. BP2 *"performs better than"* BP1. It is easy to verify that this defines a category. The transitivity of $\leq$ gives the composition relation in *Business Process* and the reflexivity guarantees the existence of the identity arrows for each object of *Business Process*. Associativity of the composition in *Business Process* obviously holds. Note that the antisymmetry property of the poset relation is not needed to verify that *Business Process* becomes a category.

The numerical relational structure is modeled as a poset $(N, \leq)$ where $\leq$ is also reflexive, transitive and antisymmetric. It can be associated with the *Measure* category where the objects are numbers (the values of the measures) and for $MV1$, $MV2$ in *Measure* there is a morphism $MV1 \rightarrow MV2$ if and only if $MV1$ and $MV2$ are in relation (such as preference), i.e. $MV1 \leq MV2$. It is easy to note that the homomorphism *Measurement Procedure: $E \rightarrow N$* between categories E and N is a functor of objects to objects along with arrows to arrows satisfying the functor property outlined above.

From the categorical perspective again we interpret a relational structure as a certain diagram of arrows visualizing the given relations between the objects which form the "nodes" of the diagram. In the following section we model the empirical structure of *Business Process* with the performance measurement and decision making process, which is the fundamental component of enterprise integration.

4.3 The Enterprise Integration Category

Putting decision-making on a firm foundation can potentially play a very important role in, for instance, risk management based on judgments on EI performance. We abstract the EI to a collection of communicating business processes and a performance control and decision making mechanism based on performance measurement.

The category *Business Process* is described in detail in [27]. The performance of each object *BP* in *Business Process* is measured (that is, *BP* is mapped to a measurement value *MV* in *Measure*) by the *Measurement Procedure* (*Measurement Procedure (BP) = MV*); each relation *BP2 "performs better than" BP1* is mapped to a relation "$\leq$" in *Measure* between the corresponding *MV1* and *MV2* values in *Measure* (*MV1 $\leq$ MV2*, or *Measurement Procedure (BP1) $\leq$ Measurement Procedure (BP2)*).

It is easy to note that the mapping *Measurement Procedure: Business Process* $\rightarrow$ *Measure* between categories *Business Process* and *Measure* is a functor of objects to objects along with arrows to arrows satisfying the functor property outlined earlier.

Figure 5 shows the performance measurement concepts by means of categorical diagram. An indicator is a measure that provides an estimate or evaluation of specified attributes derived from the analysis of the measurement data (values) with respect to defined decision criteria, which serves as basis for decision-making by the measurement users.

In order to accomplish EI compliance with the imposed performance requirements we need a distinct kind of knowledge on the EI dynamic relationships governed by decision-making rules, each rule defining its antecedents and consequents; for example, the input is performance feedback and the output are control actions required to improve the performance indicators. We model such rules as a generic functor *Decision Making Rule: Quality Feedback* $\rightarrow$ *Control Actions* which maps each object of *Quality Feedback* to and object in *Control Actions* and each preference relation in *Quality Feedback* to a criticality relation in *Control Actions*. Each control action has to be mapped to a business process; the above is modeled with the functor *Update: Control Actions* $\rightarrow$ *Business Process*.

It is to be noted that the category theory allows for different types of arrows between objects in one category. In *Business Process* category morphisms are all kind of relations between business processes. We generalize the communication between pairs of business processes as a communication arrow (see Figure 5). In our EI model we consider only one way of communication between business processes: events; this can be modeled, for instance, via "communicate" arrow between *BP1* and *BP2* in Figure 5. In an abstract sense we are dealing with arrow-diagrams of business processes where the existing arrows represent preferences or cooperation channels (communication) in a very general way. This gives us the

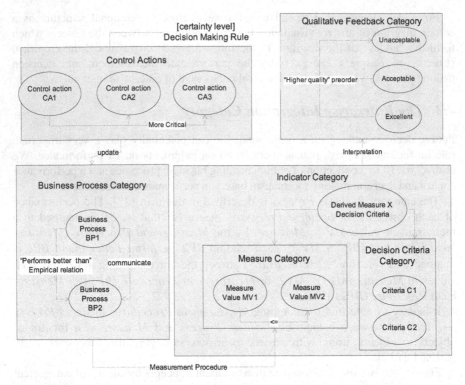

Fig. 5 Enterprise Performance measurement categorical framework

justification for associating *Business Process* category with the category *PATH* described in [32], where the morphisms are sequences (paths) of consecutive arrows. This defines composition of arrows in a natural way (concatenation of consecutive arrows) and this composition is associative. The identity arrow with respect to each object in *Business Process* will be assumed to exist by definition; graph-theoretically it is a loop is the corresponding node. This leads to the following definition: The associated *Business Process* diagram of a given EI is a category *PATH* where each node represents a business process and where the arrows represent the morphisms between the business process of preference and communication types, as described above.

One of the extensions to the model that is currently under our investigation is using Markov Decision Process for establishing the level of confidence in the choice of the decision-making rules in a given context. The concept is illustrated in the Figure 5 as the certainty level added to the decision-making rule arrow.

8 Conclusions and Future Work

The paper introduced an important direction concerning formal aspects of enterprise integration and decision-making based on performance measurement. It is motivated by the importance of enterprise integration compliance with the imposed performance requirements necessitates continuous control of the performance indicators in real time in increasingly complex large-scale enterprise integration systems.

The work exploits the applicability of category theory as a unifying formal language that allows using the same constructs for modeling heterogeneous objects and different types of relations between them. In our research, category theory provides a computational mechanism whereby such knowledge can be applied to enterprise integration data and information structures to arrive at conclusions that are valid. We have shown that category theory is expressive enough to capture knowledge about heterogeneous requirements, their interrelations and decision-making mechanism in one formal representation, where structure and reasoning are inextricably bound together.

Moreover, the categorical framework would allow decision-making based on theoretically valid measurement procedures on all levels of performance indicators: integration, enterprise, process, function, project and product.

We intend to explore the potential of an automated reasoning machine for EI based on algebra of Category Theory (algebraic approach based on a categorical calculus of relations) to measurement and decision-making. One of the important aspects that we intent to tackle concerns validation of our approach through simple scenarios of cooperating business processes.

References

1. Fenton, N.E., Pfleeger, S.L.: Software Metrics: A Rigorous and Practical Approach 2/e. PWS Publishing Company (1998)
2. Barr, M., Wells, C.: Category Theory for Computing Science. Prentice-Hall, Englewood Cliffs (1990)
3. Fiadeiro, J.: Categories for Software Engineering. Springer, Heidelberg (2005)
4. Whitmire, S.: Object Oriented Design Measurement. Whiley Computer Publishing, New York (1997)
5. Kochhar, A., Zhang, Y.: A framework for performance measurement in virtual enterprises. In: Proceedings of the 2nd International Workshop on Performance Measurement, 6-7 June, Hanover, pp. 2–11 (2002)
6. Krantz, D.H., Luce, R.D., Suppes, P., Tversky, A.: Foundations of Measurement – Additive and Polynomial Representations, vol. 1. Academic Press, New York (1971)
7. Narens, L.: Abstract Measurement Theory. MIT Press, Cambridge (1985)
8. Johnson, H.T., Kaplan, R.S.: Relevance Lost: The Rise and Fall of Management Accounting. Harvard Business School Press, Boston (1987)
9. McNair, C.J., Masconi, W.: Measuring performance in advanced manufacturing environment. Management Accounting, pp.28–31 (July 1987)
10. Kaplan, R.S.: Measures for Manufacturing Excellence. Harvard Business School Press, Boston (1990)
11. Druker, P.E.: The emerging theory of manufacturing. Harvard Business Review, 94–102 (May/June 1990)
12. Russell, R.: The role of performance measurement in manufacturing excellence. In: BPICS Conference, Birmingham (1992)

13. Cross, K.F., Lynch, R.L.: The SMART way to define and sustain success. National Productivity Review 9(1), 23–33 (1989)
14. Dixon, J.R., Nanni, A.J., Vollmann, T.E.: The New Performance Challenge – Measuring Operations for World-class Competition, Dow Jones-Irwin, Homewood, IL (1990)
15. Kaplan, R.S., Norton, D.P.: The Balanced Scorecard: Translating Strategy into Action. Harvard Business School Press, Boston (1996)
16. EFQM. Self-assessment Guidelines for Companies, European Foundation for Quality Management, Brussels (1998)
17. Beer, S.: Diagnosing the System for Organizations. Wiley, Chichester (1985)
18. Neely, A., Mills, J., Gregory, M., Richards, H., Platts, K., Bourne, M.: Getting the Measure of your Business. University of Cambridge, Cambridge (1996)
19. Neely, A., Adams, C.: The performance prism perspective. Journal of Cost Management 15(1), 7–15 (2001)
20. Capability Maturity Model Integration (CMMISM), Version 1.1, Software Engineering Institute, Pittsburgh, CMMI-SE/SW/IPPD/SS, V1.1 (March 2002)
21. Card, D.: Integrating Practical Software Measurement and the Balanced Scorecard. In: Proceedings of COMPSAC 2003 (2003)
22. Gunasekaran, A., Patel, C., Tirtiroglu, E.: Performance measures and metrics in a supply chain environment. International Journal of Operations & Production Management 21(1/2), 71–87 (2001)
23. http://www.supply-chain.org
24. Beamon, M.: Measuring supply chain performance. International Journal of Operations & Production Management 19(3), 275–292 (1999)
25. Goguen, J.A.: A categorical manifesto. Mathematical Structures in Computer Science 1, 49–67 (1991)
26. Mikhnovsky, V., Ormandjieva, O.: Towards Enterprise Integration Performance Assessment based on Category Theory. In: Proceedings of Engineering/Computing and Systems Research E-Conference CISSE 2008 (2008)
27. Ormandjieva, O., Mikhnovsky, V.: Enterprise Integration Performance Modeling and Measurement Based on Category Theory. In: Proceedings of 2009 World Congress on Computer Science and Information Engineering (CSIE 2009), Los Angeles (2009)
28. International Standard ISO/IEC 25030:2007(E), Software engineering — Software product Quality Requirements and Evaluation (SQuaRE) — Quality requirements
29. International Standard ISO/IEC 15939 Second edition (2007); Whitmire, S.: Systems and software engineering —Measurement process. In: Object Oriented Design Measurement. Whiley Computer Publishing, New York (1997)
30. Roberts, F.: Measurement Theory. Encyclopedia of Mathematics and its Applications, vol. 7. Addison-Wesley, Reading (1979)
31. Eilenberg, S., Mac Lane, S.: General Theory of Natural Equivalences. Transactions of the American Mathematical Society 58, 231–294 (1945)
32. Pfalzgraf, J.: ACCAT tutorial Presented at 27th German Conference on Artificial Intelligence (KI 2004), September 24 (2004),
    http://www.cosy.sbg.ac.at/~jpfalz/ACCAT-TutorialSKRIPT.pdf

Bayesian Statistical Modeling of System Energy Saving Effectiveness for MAC Protocols of Wireless Sensor Networks*

Myong Hee Kim and Man-Gon Park

Summary. The wireless sensor network is a wireless network consisting of spatially distributed autonomous sensor devices which are called sensor nodes in remote setting to cooperatively monitor and control physical or environmental conditions. The lifetimes of sensor nodes depend on the energy availability with energy consumption. Due to the size limitation and remoteness of sensor devices after deployment, it is not able to resupply or recharge power. The system energy saving effectiveness is the probability that the wireless sensor network system can successfully meet an energy saving operational demand. To extend the system effectiveness in energy saving, the lifetimes of sensor nodes must be increased by making them energy efficient as possible. In this paper, we propose Bayesian statistical models for observed active and sleep times data of sensor nodes under the selected energy efficient CSMA contention-based MAC protocols in consideration of the system effectiveness in energy saving in a wireless sensor network. Accordingly, we propose Bayes estimators for the system energy saving effectiveness of the wireless sensor networks by use of the Bayesian method under the conjugate prior information.

1 Introduction

The wireless sensor network (WSN) has been at the crossroads of some major information and communication technologies such as network technology, communication technology, multimedia technology, ubiquitous technology, embedded software system technology, and so on. The WSN nodes operate on battery power which is often deployed in a rough physical environment as some networks many consists of hundreds to thousands of nodes.

The media access control (MAC) protocols of WSN extend network lifetimes by reducing the activity of the highest energy-demanding component of the sensor

Myong Hee Kim and Man-Gon Park
Division of Electronic, Computer and Telecommunication Engineering,
Pukyong National University, Rep. of Korea
599-1 Daeyeon-Dong, Nam-Gu, Busan, Republic of Korea
e-mails: mhgold@naver.com, mpark@pknu.ac.kr

* This work was supported by the Pukyong National University Research Fund in 2008 (PK-2008-028).

R. Lee, N. Ishii (Eds.): Software Engineering, Artificial Intelligence, SCI 209, pp. 233–245.
springerlink.com   © Springer-Verlag Berlin Heidelberg 2009

platform. Trading off network throughput and latency, energy efficient MAC protocols synchronize network communication to create opportunities for radios to sleep with active duty cycles [1, 2, 3, 4].

In order to reduce energy consumption at sensor nodes, significant researches have been carried out on the design of low-power sensor devices. But due to fundamental hardware limitations, energy efficiency can only be achieved through the design of energy efficient communication protocols such as the sensor MAC (S-MAC), Timeout MAC (T-MAC), Berkeley MAC (B-MAC), Dynamic Sensor MAC (DS-MAC), Dynamic MAC (D-MAC) and so on [5, 6, 7, 13, 14, 15].

Recently Bayesian data analysis methods for modeling informative estimation and evaluation of network systems which is called Bayesian networks have become increasingly and rapidly important in the WSN [8, 9, 10, 11, 12, 17].

In this paper, we develop Bayesian statistical models for observed active and sleep times data based on time frames of sensor nodes under the selected energy efficient CSMA contention-based MAC protocols such as S-MAC, T-MAC, B-MAC, DS-MAC, and D-MAC based on IEEE 802.15.4 for wireless sensor networks. Also we propose Bayesian estimation procedures for system effectiveness in energy saving for the selected energy efficient MAC protocols under the consideration of the system effectiveness in energy saving based on active and sleep times data from sensor nodes in a wireless sensor network. Accordingly, we propose Bayes estimators for the system effectiveness in energy saving of the wireless sensor networks by use of the Bayesian method under the conjugate prior information about means of active and sleep times based on time frames of sensor nodes in a wireless sensor network.

2 Bayesian Statistical Modeling of System Energy Saving Effectiveness under the Conjugate Priors

We consider a deployed wireless sensor network system for continuous sensing, event detection, and monitoring consisting of N non-identical sensor nodes each of which has exponentially distributed active and sleep times under the lifetime frames of a wireless sensor network according to MAC protocols at the data-link layer.

More precisely, for i^{th} sensor node, if we suppose that k_i active/sleep frame cycles (slots) are observed as lifetimes, then $A_{i1}, A_{i2}, \cdots, A_{ik_i}$ are independent active times random sample data from the exponential distribution $\varepsilon(\theta_i)$ with the mean time between active times (MTBA) θ_i and $S_{i1}, S_{i2}, \cdots, S_{ik_i}$ are independent sleep times random sample data from the exponential distribution $\varepsilon(\mu_i)$ with the mean time between sleep times (MTBS) μ_i, and the component energy saving effectiveness of the ith sensor node is given by

$$Q_i = \frac{\mu_i}{\theta_i + \mu_i}.$$

(1)

According to Sandler [15], the system energy saving un-effectiveness of a wireless sensor network is the product of the component energy saving un-effectiveness of the ith sensor node, that is,

$$\overline{Q} = \prod_{i=1}^{N} \overline{Q}_i = \prod_{i=1}^{N}(1 - \frac{\mu_i}{\theta_i + \mu_i}) = \prod_{i=1}^{N}\frac{\theta_i}{\theta_i + \mu_i}. \tag{2}$$

Therefore, the system effectiveness in energy saving of a wireless sensor network which are consisting of N non-identical sensor nodes becomes

$$Q = 1 - \overline{Q} = 1 - \prod_{i=1}^{N}\frac{\theta_i}{\theta_i + \mu_i}, \tag{3}$$

where θ_i is MTBA and μ_i is MTBS of the ith sensor node.

The fundamental tool used for Bayesian estimation is Bayes' theorem. The likelihood function is the function through which the observed lifetimes results or samples (active and sleep times) from the deployed sensor nodes modify prior information of the system effectiveness. The prior distribution represents all information that is known or assumed about the system effectiveness in energy saving. The posterior distribution is a modified and updated version of the previous information expressed by the prior distribution on the basis of the observed lifetimes results or samples from the deployed sensor nodes of a wireless sensor network.

We may write Bayes' theorem in words as Posterior distribution = Prior distribution x Likelihood Function / Marginal Distribution. That is,

$$f(\hat{\theta} \mid \underline{x}) = \frac{[\prod_{i=1}^{N}(f(x_i \mid \hat{\theta})]g(\hat{\theta})}{f(x)}$$

$$= \frac{f(\underline{x} \mid \hat{\theta})g(\hat{\theta})}{f(\underline{x})} = \frac{g(\hat{\theta})L(\hat{\theta} \mid \underline{x})}{\int_{\hat{\Theta}} g(\hat{\theta})L(\hat{\theta} \mid \underline{x})d\hat{\theta}}, \text{ if } \hat{\theta} \text{ is continuous,} \tag{4}$$

where, $\hat{\theta} = (\theta_1, \theta_2, \cdots, \theta_k)$ is a vector of the parameter space $\hat{\Theta}$, $\underline{x} = (x_1, x_2, \cdots, x_N)$ is a vector of statistically independent observation of the random variables $\underline{X}$ (the sample data), $g(\hat{\theta})$ is the joint prior probability distribution of $\hat{\Theta}$ (the prior knowledge), $f(x_i \mid \hat{\theta})$ is the conditional probability distribution of x_i given $\hat{\theta}$ (the sampling model), $f(\underline{x} \mid \hat{\theta}) = \prod_{i=1}^{N} f(x_i \mid \hat{\theta}) = L(\hat{\theta} \mid \underline{x})$ is the joint conditional probability distribution of $\underline{X}$ given $\hat{\theta}$ (the likelihood function of $\hat{\theta}$ give $\underline{x}$), $f(\underline{x}, \hat{\theta})$ is the joint probability distribution of $\underline{X}$ and $\hat{\Theta}$, $f(x)$ is the marginal probability distribution of $\underline{X}$, and $f(\hat{\theta} \mid \underline{x})$ is the joint posterior probability distribution of $\hat{\theta}$ given $\underline{x}$ (the posterior model).

If we assume that the active times { $A_{i1}, A_{i2}, \cdots, A_{ik_i}$ } for the N non-identical sensor nodes are exponentially distributed with Mean Times Between Actives (MTBA's) $\theta_1, \theta_2, \cdots, \theta_N$, respectively, such that

$$f_1(A_{ij} \mid \theta_i) = \frac{1}{\theta_i} \exp(-\frac{A_{ij}}{\theta_i}), \tag{5}$$

where, $A_{ij} (> 0)$ is the jth active time of the ith sensor node for $j = 1, \cdots, k_i$, $\theta_i (> 0)$ is the Mean Time Between Actives (MTBA) of the ith sensor node, $k_i =$ number of the observed active/sleep cycles (slots) of the ith sensor node, for $i = 1, \cdots, N$.

Also if we assume that the sleep times $\{ S_{i1}, S_{i2}, \cdots, S_{ik_i} \}$ for the N non-identical sensor nodes are exponentially distributed with Mean Times Between Sleeps (MTBS's) $\mu_1, \mu_2, \cdots, \mu_N$ respectively, such that

$$f_2(S_{ij} \mid \mu_i) = \frac{1}{\mu_i} \exp(-\frac{S_{ij}}{\mu_i}), \tag{6}$$

where, $S_{ij} (> 0)$ is the jth sleep time of the ith sensor node, for $j = 1, \cdots, k_i$, $\mu_i (> 0)$ is the Mean Time Between Sleeps (MTBS) of the ith sensor node, and k_i is the number of the observed active/sleep cycles (slots) of the ith sensor node, for $i = 1, \cdots, N$.

For i^{th} sensor node, we obtain the likelihood function of T_{A_i} and T_{S_i} given θ_i and μ_i as follows.

$$L_i(\theta_i, \mu_i \mid T_{A_i}, T_{S_i}) = \prod_{j=1}^{k_i} f_1(A_{ij} \mid \theta_i) f_2(S_{ij} \mid \mu_i)$$

$$= \prod_{j=1}^{k_i} \frac{1}{\theta_i} \exp(-\frac{A_{ij}}{\theta_i}) \frac{1}{\mu_i} \exp(-\frac{S_{ij}}{\mu_i})$$

$$= \frac{1}{(\theta_i \mu_i)^{k_i}} \exp[-(\frac{T_{A_i}}{\theta_i} + \frac{T_{S_i}}{\mu_i})], \tag{7}$$

where,

k_i = number of the observed active/sleep cycles (slots) of the ith sensor node,

T_{A_i} = total operating time observed; $T_{A_i} = \sum_{j=1}^{k_i} A_{ij}$, A_{ij} is the jth active time of

the ith sensor node, and

T_{S_i} = total sleep time observed; $T_{S_i} = \sum_{j=1}^{k_i} S_{ij}$, S_{ij} is the jth sleep time of the

i^{th} sensor node,

for $i = 1, \cdots, N$.

The prior distribution as a prior knowledge of parameters can be chosen to represent the beliefs or experience of the researchers or engineers before observing the results or samples of an experiment. However, it is hard for a researchers or engineers to specify prior beliefs or experience about parameters, and to cast them into the form of a prior probability distribution.

A well-known strategy is to select a prior with a suitable form so the posterior distribution belongs to the same family as the prior distribution. The choice of the family depends on the likelihood. The prior and posterior distributions chosen by this way are said to be conjugate. This prior distribution is called as conjugate prior.

More precisely, if $\hat{\theta} = (\theta_1, \theta_2, \cdots, \theta_k)$ is a vector of the parameter space $\hat{\Theta}$, a class of prior probability distributions $g(\hat{\theta})$ is said to be conjugate to a class of likelihood functions $L(\hat{\theta} \mid \underline{x})$ if the resulting posterior distributions $f(\hat{\theta} \mid \underline{x})$ are in the same family as $g(\hat{\theta})$. Such a choice is a conjugate prior. For example, the Gaussian family is conjugate to itself (or self-conjugate). If the likelihood function is Gamma, choosing a Gamma prior will ensure that the posterior distribution is also Gamma. The concept as well as the term "conjugate prior" was well introduced by Howard Raiffa and Robert Schlaifer [18] and De Groot [19] in their works on Bayesian decision theory. Therefore the conjugate prior for the exponential distribution is the gamma distribution.

Thus, for the MTBA θ_i and MTBS μ_i in the ith sensor node, two general classes of the conjugate prior distributions of θ_i and μ_i are given by

$$g_i(\theta_i) = \frac{b_i^{a_i}}{\Gamma(a_i)} \frac{1}{\theta_i^{a_i+1}} \exp(-\frac{b_i}{\theta_i}), \ \theta_i > 0, \ a_i, b_i > 0, \tag{8}$$

and

$$g_i(\mu_i) = \frac{d_i^{c_i}}{\Gamma(c_i)} \frac{1}{\mu_i^{c_i+1}} \exp(-\frac{d_i}{\mu_i}), \ \mu_i > 0, c_i, d_i > 0, \tag{9}$$

respectively.

From the conjugate prior distributions in (8) and (9), and the likelihood function in (7), we can calculate joint posterior distribution of θ_i and μ_i for the ith sensor node as follows.

$$\overline{f}_i(\theta_i, \mu_i \mid T_{A_i}, T_{S_i})$$

$$= \frac{L_i(\theta_i, \mu_i \mid T_{A_i}, T_{S_i})}{\int_0^\infty \int_0^\infty L_i(\theta_i, \mu_i \mid T_{A_i}, T_{S_i})} \cdot \frac{g_i(\theta_i) g_i(\mu_i)}{g_i(\theta_i) g_i(\mu_i) d\theta_i d\mu_i}$$

$$= \frac{(\theta_i \mu_i)^{-k_i} \exp[-(\frac{T_{A_i}}{\theta_i} + \frac{T_{S_i}}{\mu_i})] \quad \theta_i^{-(a_i+1)} \exp(-\frac{b_i}{\theta_i}) \cdot \mu_i^{-(c_i+1)} \exp(-\frac{d_i}{\mu_i})}{\int_0^\infty \int_0^\infty (\theta_i \mu_i)^{-k_i} \exp[-(\frac{T_{A_i}}{\theta_i} + \frac{T_{S_i}}{\mu_i})] \ \theta_i^{-(a_i+1)} \exp(-\frac{b_i}{\theta_i}) \cdot \mu_i^{-(c_i+1)} \exp(-\frac{d_i}{\mu_i}) d\theta_i d\mu_i}$$

$$
\theta_i^{-(k_i+a_i+1)} \mu_i^{-(k_i+c_i+1)} \exp[-(\frac{T_{A_i}+b_i}{\theta_i}+\frac{T_{S_i}+d_i}{\mu_i})]
$$

$$
= \frac{}{\int_0^\infty \int_0^\infty \theta_i^{-(k_i+a_i+1)} \cdot \mu_i^{-(k_i+c_i+1)} \exp[-(\frac{T_{A_i}+b_i}{\theta_i}+\frac{T_{S_i}+d_i}{\mu_i})] d\theta_i d\mu_i}
$$

$$
= \frac{(T_{A_i}+b_i)^{k_i+a_i}(T_{S_i}+d_i)^{k_i+c_i}}{\Gamma(k_i+a_i)\Gamma(k_i+c_i)} \cdot \frac{\exp[-(\frac{T_{A_i}+b_i}{\theta_i}+\frac{T_{S_i}+d_i}{\mu_i})]}{\theta_i^{k_i+a_i+1}\mu_i^{k_i+c_i+1}}, \tag{10}
$$

where $\theta_i > 0$, $\mu_i > 0$ and $\Gamma(\cdot)$ is a gamma function.

3 Bayesian Estimation Procedure of the System Energy Saving Effectiveness

Under squared-error loss and the conjugate prior knowledge, the Bayes point estimator of the component energy saving un-effectiveness of ith deployed sensor node is posterior mean of $\overline{Q}_i$.

First, we have to find the posterior distribution of $\delta_i = \mu_i / \theta_i$. According to (10), we can express that

$$
\overline{g}_{\delta_i}(\delta_i \mid T_{A_i},T_{S_i}) = \int_0^\infty \overline{f_i}(\theta_i,\theta_i\delta_i \mid T_{A_i},T_{S_i}) \cdot \theta_i d\theta_i
$$

$$
= \frac{(T_{A_i}+b_i)^{k_i+a_i}(T_{S_i}+d_i)^{k_i+c_i}}{\Gamma(k_i+a_i)\Gamma(k_i+c_i)} \cdot \int_0^\infty \theta_i^{-(k_i+a_i)}(\theta_i\delta_i)^{-(k_i+c_i+1)} \exp[-(\frac{T_{A_i}+b_i}{\theta_i}+\frac{T_{S_i}+d_i}{\theta_i\delta_i})]d\theta_i
$$

$$
= \frac{(T_{A_i}+b_i)^{k_i+a_i}(T_{S_i}+d_i)^{k_i+c_i}\delta_i^{-(k_i+c_i+1)}}{\Gamma(k_i+a_i)\Gamma(k_i+c_i)} \cdot \int_0^\infty \theta_i^{-(2k_i+a_i+c_i+1)} \exp[-\frac{1}{\theta_i}(\frac{T_{S_i}+d_i}{\delta_i}+T_{A_i}+b_i)]d\theta_i. \tag{11}
$$

Substituting $z = 1/\theta_i$ and using the formula $\int_0^\infty z^{a-1}\exp(-tz)dz = t^{-a}\Gamma(a)$, we can obtain the posterior distribution of $\delta_i = \mu_i / \theta_i$.

$$
\overline{g}_{\delta_i}(\delta_i \mid T_{A_i},T_{S_i})
$$

$$
= \frac{(T_{A_i}+b_i)^{k_i+a_i}(T_{S_i}+d_i)^{k_i+c_i}\delta_i^{-(k_i+c_i+1)}}{\Gamma(k_i+a_i)\Gamma(k_i+c_i)} \cdot \int_0^\infty z^{(2k_i+a_i+c_i)-1} \exp[-(\frac{T_{S_i}+d_i}{\delta_i}+T_{A_i}+b_i)z]dz
$$

$$= \frac{(T_{A_i} + b_i)^{k_i+a_i}(T_{S_i} + d_i)^{k_i+c_i} \delta_i^{-(k_i+c_i-1)}}{\Gamma(k_i+a_i)\Gamma(k_i+c_i)} \cdot \frac{\Gamma(2k_i+a_i+c_i)}{(\dfrac{T_{S_i}+d_i}{\delta_i} + T_{A_i} + b_i)^{2k_i+a_i+c_i}}$$

$$= \frac{(T_{A_i} + b_i)^{k_i+a_i}(T_{S_i} + d_i)^{k_i+c_i} \delta_i^{k_i+a_i-1}}{\Gamma(k_i+a_i)\Gamma(k_i+c_i)} \cdot \frac{\Gamma(2k_i+a_i+c_i)}{[T_{S_i}+d_i + \delta_i(T_{A_i}+b_i)]^{2k_i+a_i+c_i}}$$

$$= \frac{(T_{A_i} + b_i)^{k_i+a_i}(T_{S_i} + d_i)^{k_i+c_i}}{B(k_i+a_i, k_i+c_i)} \cdot \frac{\delta_i^{k_i+a_i-1}}{[T_{S_i}+d_i + \delta_i(T_{A_i}+b_i)]^{2k_i+a_i+c_i}}. \quad (12)$$

Hence, we can calculate the posterior distribution of the component energy saving un-effectiveness $\overline{Q}_i$ for i^{th} sensor node under the conjugate priors as follows.

$$\overline{g}_i(\overline{Q}_i \mid T_{A_i}, T_{S_i}) = \overline{g}_{\delta_i}(\overline{Q}_i^{-1} - 1 \mid T_{A_i}, T_{S_i})\overline{Q}_i^{-2}$$

$$= \frac{(T_{A_i} + b_i)^{k_i+a_i}(T_{S_i} + d_i)^{k_i+c_i}}{B(k_i+a_i, k_i+c_i)} \cdot \frac{(\overline{Q}_i^{-1} - 1)^{k_i+a_i-1}\overline{Q}_i^{-2}}{[T_{S_i}+d_i + (\overline{Q}_i^{-1}-1)(T_{A_i}+b_i)]^{2k_i+a_i+c_i}}$$

$$= \frac{[\dfrac{T_{S_i}+d_i}{T_{A_i}+b_i}]^{k_i+c_i}}{B(k_i+a_i, k_i+c_i)} \cdot \frac{(1-\overline{Q}_i)^{k_i+a_i-1}(\overline{Q}_i)^{k_i+c_i-1}}{[1-\overline{Q}_i(1-\dfrac{T_{S_i}+d_i}{T_{A_i}+b_i})]^{2k_i+a_i+c_i}}$$

$$= \frac{[\dfrac{T_{S_i}+d_i}{T_{A_i}+b_i}]^{k_i+c_i}}{B(k_i+a_i, k_i+c_i)} \cdot (\overline{Q}_i)^{k_i+c_i-1}(1-\overline{Q}_i)^{k_i+a_i-1}[1-\overline{Q}_i(1-\dfrac{T_{S_i}+d_i}{T_{A_i}+b_i})]^{-(2k_i+a_i-c_i)}, \quad (13)$$

where $0 < \overline{Q}_i = \dfrac{1}{1+\delta_i} < 1$, $B(\cdot,\cdot)$ is a beta function and $\delta_i = \mu_i / \theta_i$ is the service factor.

Under the squared-error loss and the conjugate priors, Bayes point estimator of the ith component energy saving un-effectiveness $\overline{Q}_i$ is the posterior mean of $\overline{Q}_i$. We can calculate the Bayes point estimator of the ith component energy saving un-effectiveness $\overline{Q}_i$ from the expression (13) and by use of transformation,

$$w = \frac{\overline{Q}_i[\dfrac{T_{S_i}+d_i}{T_{A_i}+b_i}]}{1 - \overline{Q}_i[1 - \dfrac{T_{S_i}+d_i}{T_{A_i}+b_i}]}. \quad (14)$$

Hence, the Bayes estimator $\overline{Q}_i^{BE}$ of the ith component energy saving un-effectiveness $\overline{Q}_i$ under the conjugate priors can be calculated by

$$\overline{Q}_i^{BE}$$

$$= E(\overline{Q}_i \mid T_{A_i}, T_{S_i})$$

$$= \int_0^1 \overline{Q}_i \overline{g}_i(\overline{Q}_i \mid T_{A_i}, T_{S_i}) d\overline{Q}_i$$

$$= \int_0^1 \frac{(\frac{T_{S_i}+d_i}{T_{A_i}+b_i})^{k_i+c_i}}{B(k_i+a_i,k_i+c_i)} \cdot \frac{(1-\overline{Q}_i)^{k_i+a_i-1}(\overline{Q}_i)^{k_i+c_i}}{[1-\overline{Q}_i(1-\frac{T_{S_i}+d_i}{T_{A_i}+b_i})]^{2k_i+a_i+c_i}} d\overline{Q}_i$$

$$= \frac{(\frac{T_{S_i}+d_i}{T_{A_i}+b_i})^{k_i+c_i}}{B(k_i+a_i,k_i+c_i)} \cdot (\frac{T_{A_i}+b_i}{T_{S_i}+d_i})^{k_i+c_i} (\frac{T_{A_i}+b_i}{T_{S_i}+d_i}) \cdot \int_0^1 w^{k_i+c_i} \cdot (1-w)^{k_i+a_i-1} \cdot [1-w(1-\frac{T_{A_i}+b_i}{T_{S_i}+d_i})]^{-1} dw$$

$$= \frac{(\frac{T_{S_i}+d_i}{T_{A_i}+b_i})}{B(k_i+a_i,k_i+c_i)} \cdot \int_0^1 w^{k_i+c_i}(1-w)^{k_i+a_i-1}[1-w(1-\frac{T_{A_i}+b_i}{T_{S_i}+d_i})]^{-1} dw$$

$$= \frac{T_{A_i}+b_i}{T_{S_i}+d_i} \cdot \frac{k_i+c_i}{2k_i+a_i+c_i} {}_2F_1(1,k_i+c_i+1;2k_i+a_i+c_i+1;1-\frac{T_{A_i}+b_i}{T_{S_i}+d_i}), \quad (15)$$

where $0 < \frac{T_{A_i}+b_i}{T_{S_i}+d_i} < 2$ and ${}_2F_1(a,b;c;t) = \sum_{i=0}^{\infty} \frac{(a)_i(b)_i}{(c)_i i!} t^i$ is a confluent hyper-geometric function in Gauss' form for $\mid t \mid < 1$ with $(a)_i = \frac{\Gamma(a+i)}{\Gamma(a)}$, $a,b,c > 0$.

Therefore, under the conjugate prior knowledge, the Bayes point estimator Q^{BE} of the system effectiveness in energy saving of the deployed sensor nodes of a wireless sensor network by means of MAC protocols can be represented as

$$Q^{BE}$$

$$= 1 - \prod_{i=1}^{N} \overline{Q}_i^{BE}$$

$$= 1 - \prod_{i=1}^{N} [\frac{T_{A_i}+b_i}{T_{S_i}+d_i} \cdot \frac{k_i+c_i}{2k_i+a_i+c_i} {}_2F_1(1,k_i+c_i+1;2k_i+a_i+c_i+1;1-\frac{T_{A_i}+b_i}{T_{S_i}+d_i})], \quad (16)$$

where $0 < \dfrac{T_{A_i} + b_i}{T_{S_i} + d_i} < 2$ and $_2F_1(a,b;c;t) = \sum_{i=0}^{\infty} \dfrac{(a)_i(b)_i}{(c)_i i!} t^i$ is a confluent hyper-

geometric function in Gauss' form for $|t| < 1$ with $(a)_i = \dfrac{\Gamma(a+i)}{\Gamma(a)}$, $a,b,c > 0$.

4 Case Study of the Proposed Bayesian Estimators

We can generate active and sleep times of 5 sensor nodes as shown in **Table 1** according to the contention-based slotted MAC protocols assuming that duty cycles (slots/frames) $k_i = 5$, message interval=0.1, MTBA=12 *msec* (milliseconds), and MTBS=20 *msec*.

With theses generated sample, we calculate Bayes estimates of the system effectiveness in energy saving according to parameters of conjugate prior knowledge

Table 1 A generated random sample of active and sleep times

(Unit: *msec*)

MAC	Node	A	B	C	D	E
802.11	Active times	12.0, 12.0, 12.0, 12.0, 12.0	12.0, 12.0, 12.0, 12.0, 12.0	12.0, 12.0, 12.0, 12.0, 12.0	12.0, 12.0, 12.0, 12.0, 12.0	12.0, 12.0, 12.0, 12.0, 12.0
	Sleep times	20.0, 20.0, 20.0, 20.0, 20.0	20.0, 20.0, 20.0, 20.0, 20.0	20.0, 20.0, 20.0, 20.0, 20.0	20.0, 20.0, 20.0, 20.0, 20.0	20.0, 20.0, 20.0, 20.0, 20.0
S-MAC	Active times	3.1, 1.2, 4.5, 2.1, 1.2	2.3, 2.1, 1.7, 1.5, 3.4	2.7, 3.6, 5.1, 3.2, 1,7	6.7, 2.9, 3.7, 2.8, 4.5	1.7, 3.2, 4.2, 2.8, 4.1
	Sleep times	28.9, 31.8, 27.5, 29.9, 31.8	29.7, 29.9, 30.3, 31.5, 28.6	29.3, 28.4, 27.9, 28.6, 30.3	25.3, 29.1, 28.3, 29.2, 27.5	30.3, 28.8, 27.8, 29.2, 27.9
DS-MAC	Active times	8.1, 10.3, 9.2, 7.5, 12.0	10.3, 4.2, 7.7, 9.5, 11.6	8.5, 10.0, 10.2, 11.3, 9.7	9.0, 7.8, 10.5, 9.4, 7.8	6.8, 7.5, 4.9, 6.9, 9.7
	Sleep times	13.9, 11.7, 12.8, 14.5, 10.0	11.7, 17.8, 14.3, 12.5, 10.4	14.6, 13.5, 12.0, 11.8, 12.3	13.0, 14.2, 11.5, 12.6, 14.2	15.2, 14.5, 17.1, 15.1, 12.3
T-MAC	Active times	1.2, 2.3, 0.9, 0.7, 1.3	0.7, 2.1, 1.6, 0.8, 1.5	2.1, 1.4, 0.7, 1.2, 1.8	0.9, 1.7, 1.9, 2.0, 1.3	1.5, 1.8, 0.7, 1.9, 2.4
	Sleep times	30.8, 29.7, 31.1, 31.3, 30.7	31.3, 29.9, 30.4, 31.2, 30.5	29.9, 30.6, 31.3, 30.8, 30.2	31.1, 30.3, 30.1, 30.0, 31.7	30.5, 30.2, 31.3, 30.1, 29.6

$a_i = 1(2)5$, $b_i = 1(2)5$, $c_i = 1(2)5$, $d_i = 1(2)5$, duty cycles $k_i = 5$ message interval=0.1(0.2)0.9, MTBA=12 *msec* and MTBS=20 *msec*. And we compare the Bayes estimates of system energy saving effectiveness under the conjugate prior knowledge and selected slotted contention-based energy efficient MAC protocols such as 802.11, S-MAC, DS-MAC, and T-MAC.

System Energy Saving Effectiveness

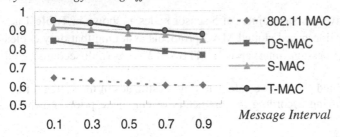

Fig. 1 Comparison of the Bayes estimates of system energy saving effectiveness with conjugate prior information $(a_i = 1, b_i = 1, c_i = 1, d_i = 1)$

System Energy Saving Effectiveness

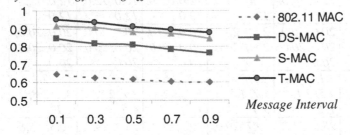

Fig. 2 Comparison of the Bayes estimates of system energy saving effectiveness with conjugate prior information $(a_i = 1, b_i = 1, c_i = 3, d_i = 3)$

System Energy Saving Effectiveness

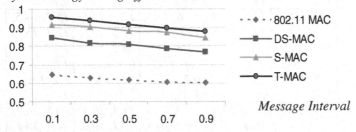

Fig. 3 Comparison of the Bayes estimates of system energy saving effectiveness with conjugate prior information $(a_i = 3, b_i = 3, c_i = 1, d_i = 1)$

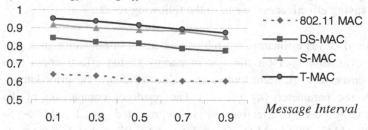

Fig. 4. Comparison of the Bayes estimates of system energy saving effectiveness with conjugate prior information $(a_i = 3, b_i = 3, c_i = 3, d_i = 3)$

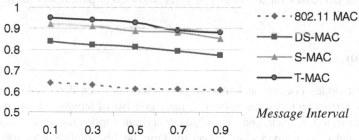

Fig. 5 Comparison of the Bayes estimates of system energy saving effectiveness with conjugate prior information $(a_i = 5, b_i = 5, c_i = 1, d_i = 1)$

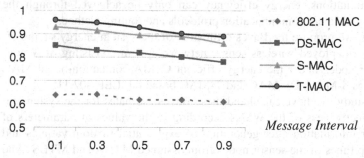

Fig. 6 Comparison of the Bayes estimates of system energy saving effectiveness with conjugate prior information $(a_i = 5, b_i = 5, c_i = 5, d_i = 5)$

From the figures (Fig. 1-6) for the comparison of the Bayes estimates of the system energy saving effectiveness, we have the following results:

(1) The proposed Bayes estimators are robust and stable in utilization as a performance evaluation tool for the system energy saving effectiveness in a wireless sensor network according to the values of conjugate prior knowledge with the parameters (a_i, b_i, c_i, d_i) The result of comparison of the system energy saving effectiveness is represented as T-MAC $\succ$ S-MAC $\succ$ DS-MAC $\succ$ 802.11 MAC, if we define "A $\succ$ B" as that A is more effective than B, for $a_i = 1(2)5, b_i = 1(2)5, c_i = 1(2)3, d_i = 1(2)3$ duty cycles (slots/frames) $k_i = 5$ message interval = 0.1(0.2)0.9, MTBA=12 $msec$ and MTBS=20 $msec$.

(2) When message interval increases from 0.1 to 0.9, the system energy saving effectiveness decreases for all values of prior knowledge parameters as we expected.

5 Conclusion

The WSN sensor nodes operate on battery power which is often deployed in a rough physical environment as some networks many consists of hundreds to thousands of nodes to monitor, track, and control many civilian application areas such as environment and habitat monitoring, object tracking, healthcare, fire detection, traffic management, home automation, and so on.

In WSN sensor nodes, MAC synchronizes the channel access in an environment where numerous sensor nodes access a shared communication medium. In order to reduce energy consumption at sensor nodes, significant researches have been carried out on the design of low-power sensor devices. But due to fundamental hardware limitations, energy efficiency can only be achieved through the design of energy efficient communication protocols and routing methods.

In this paper, we propose the Bayes estimators of the system energy saving effectiveness for a deployed wireless sensor network system consisting of N nonidentical sensor nodes under the energy efficient CSMA contention-based MAC protocols such as S-MAC, DS-MAC and T-MAC based on IEEE 802.11.

In the case study, we have calculated these Bayes estimates for the system energy saving effectiveness of the WSNs according to the values of parameters of conjugate prior information after generation of exponential random variates and active and sleep times of the sensor nodes from assumed MTBA and MTBS. And we have compared the system effectiveness in energy saving according to the slotted contention-based energy efficient MAC protocols such as 802.11, S-MAC, DS-MAC, and T-MAC. And we have recognized that the proposed Bayesian system energy saving effectiveness estimation tools are excellent to adapt in evaluation of energy efficient contention-based MAC protocols using prior knowledge from previous experience.

References

1. Bharghavan, V., Demers, A., Shenker, S., Zhang, L.: MACAW: A Media Access Proto col for Wireless LAN's. In: Proceedings of the ACM SIGCOMM Conference on Com munications Architectures, Protocols and Applications, London, UK, 31 August- 2 Sep tember, pp. 212–225 (1994)
2. Demirkol, I., Ersoy, C., Alagöz, F.: MAC Protocols for Wireless Sensor Networks: A Survey. IEEE Communications Magazine 44(4), 115–121 (2006)
3. Halkes, G.P., van Dam, T., Langendoen, K.G.: Comparing energy-saving MAC protoc ols for wireless sensor networks. In: Mobile Networks and Applications, vol. 10(5), pp . 783–791. Kluwer Academic Publishers, Hingham (2005)
4. van Dam, T., Langendoen, K.: An Adaptive Energy-Efficient MAC Protocol for Wireles s Sensor Networks. In: Proceedings of the 1st International Conference on Embedded Net worked Sensor Systems, Los Angeles, USA, November 5-7, 2003, pp. 171–180 (2003)
5. Polastre, J., Culler, D.: B-MAC: An Adaptive CSMA Layer for Low-Power Operation. Technical Report CS294-F03/BMAC, UC Berkeley (December 2003)
6. Sagduyu, Ephremides, Y.E.: The Problem of Medium Access Control in Wireless Sens or Networks. IEEE Wireless Communications 1(6), 44–53 (2004)
7. Dunkels, A., Österlind, F., Tsiftes, N., He, Z.: Software-based sensor node energy esti mation. In: Proceedings of the 5th International Conference on Embedded Networked Sensor Systems, Sydney, Australia, November 6-9, 2007, pp. 409–410 (2007)
8. Stemm, M., Katz, R.H.: Measuring and Reducing Energy Consumption of Network Int erfaces in Hand-held Devices. IEICE Transactions on Communications E80-B, 1125–1 131 (1997)
9. Pourret, O., Naïm, P., Marcot, B.: Bayesian Networks: A Practical Guide to Applicatio ns. John Wiley & Sons Ltd., Chichester (2008)
10. Neil, M., Fenton, N.E., Tailor, M.: Using Bayesian Networks to model Expected and U nexpected Operational Losses. An International Journal of Risk Analysis 25(4), 963–9 72 (2005)
11. Murphy, K.P.: Dynamic Bayesian Networks: Representation, Inference and Learning. Ph.D. thesis, UC Berkeley, Computer Science Division (July 2002)
12. Stann, F., Heidemann, J.: BARD: Bayesian-Assisted Resource Discovery in Sensor Ne tworks. In: Proceedings of the. 24th Annual Joint Conference of the IEEE Computer a nd Communications Societies (INFOCOM 2005), Miami, Florida, USA, March 13-17, 2005, vol. 2, pp. 866–877 (2005)
13. Martz, H.F., Waller, R.A.: Bayesian Reliability Analysis of Complex Series/Parallel Syst ems of Binomial Subsystems and Components. Technometrics 32(4), 407–416 (1990)
14. Gutierrez, J.A., Naeve, M., Callaway, E., Bourgeois, M., Mitter, V., Heile, B.: IEEE 8 02.15.4: A Developing Standard for Low-Power Low-Cost Wireless Personal Area Net works. IEEE Network 15(5), 12–19 (2001)
15. IEEE 802.15 WPAN^{TM} Task Group 4 (TG4), http://www.ieee802.org/15/ pub/TG4.html (retrieved, 7th March, 2009)
16. Sandler, G.H.: System Reliability Engineering. Prentice-Hall, Englewood Cliffs (1963)
17. Pearl, J.: Bayesian Networks: A Model of Self-Activated Memory for Evidential Reas oning. In: Proceedings of the 7th Conference of the Cognitive Science Society, August 15-17, 1985, pp. 329–334. University of California, Irvine (1985)
18. Raiffa, H., Schlaifer, R.: Applied Statistical Decision Theory (1st edn. Harvard Univer sity Press, Cambridge, 1961), Paperback edn. John Wiley & Sons, Chichester (2000)
19. De Groot, M.H.: Optimal Statistical Decisions. McGraw-Hill, New York (1970)

Field Server System Using Solar Energy Based on Wireless Sensor

Chang-Sun Shin, Su-Chong Joo, Yong-Woong Lee, Choun-Bo Sim, and Hyun Yoe

Summary. In agriculture and livestock industry, environments, like temperature, humidity, CO2, and water supply, are important factors which decide on a growth speed, an output, taste, and etc. If you monitor and manage the above environments efficiently, you can get good results in products. For monitoring and managing the environments, this paper proposes a Ubiquitous Field Server System (UFSS) on wireless sensor networks. Also this system uses solar energy. The UFSS can monitor and collect the information of field environments and the system's location using the environment and soil sensors, CCTV camera, GPS (Global Positioning System) module, and solar cell module without restriction of the power supply and the system's location. This system composes of three layers. The devices layer includes sensors, GPS, CCTV camera, and solar cell. The middle layer consists of the soil manager, the location manager, the motion manager, the information storage, and the web server. The application layer provides with the soil and environments monitoring service, the location monitoring service, the motion monitoring service, and statistics service. Finally, we apply the each layer's modules to the UFSS, and show the executing results of our system using GUIs.

1 Introduction

With advent of a large-scale agricultural and livestock field, we need to monitor the climate and environment information at a remote place. In networks and IT technologies, the physical user space and the cyber computing space are blended. Ubiquitous computing is new paradigm that can be provided human with the dynamic configured services according to changes of surrounding environments [1,2]. If we apply Ubiquitous IT technologies to this field, there can be improved

Chang-Sun Shin, Yong-Woong Lee, Choun-Bo Sim, and Hyun Yoe
School of Information and Communication Engineering,
Sunchon National University, Korea
e-mail: {csshin, ywlee, cbsim, yhyun}@sunchon.ac.kr

Su-Chong Joo
School of Electrical, Electronic and Information Engineering,
Wonkwang University, Korea
e-mail: scjoo@wonkwang.ac.kr

R. Lee, N. Ishii (Eds.): Software Engineering, Artificial Intelligence, SCI 209, pp. 247–254.
springerlink.com   © Springer-Verlag Berlin Heidelberg 2009

their outputs. Also we monitor and manage climate or environments efficiently [3,4,5].

This paper proposes a Ubiquitous Field Server System (UFSS) that monitors and collects the information of field environments and the system's location, and equips with a self-power supply.

2 Related Works

Light, air, water, temperature, and soil are the essential elements of agriculture. The Jet Propulsion Lab. in NASA studied the solar environment sensors to monitor temperature, humidity and oxygen of environment and soil. They applied the Sensor Web 3.1 with low power and small size.

The Phytalk Co. in Israel developed a plant growth monitoring system. This system measures the environment status by using sensors adhered to plant and sends information to farmer's home via internet.

Also agricultural sensors and robots are proposed, and various applications are developed.

3 Ubiquitous Field Server System (UFSS)

The UFSS has three layers. The device layer includes sensors, GPS, CCTV camera, and solar cell. The middle layer has the soil manager, the location manager, the motion manager, the information storage, and the web server. The application layer provides with the soil and environments monitoring service, the location monitoring service, and the motion monitoring service. These three layers are integrated into the UFSS. By interacting with each layer, the system provides us with field environment information. Figure 1 shows the architecture of the UFSS.

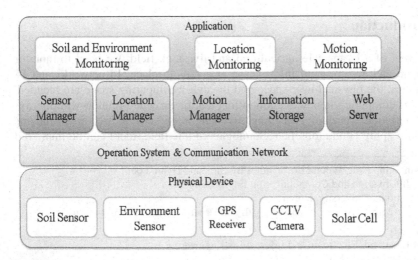

Fig. 1 UFSS architecture

Stream		EnvData		GPSdata		SoilData	
PK	seq	PK	seq ·	PK	seq	PK	seq
	Video Date		Id Temp Humi Light Date		Longitude Latitude Time Date		Id Temp Humi Ec Batt Date

Fig. 2 Database schema in the information storage

Let's introduce components of middle layer. The sensor manager manages the information from soil sensor and environment sensor. The location manager interacts with GPS module in the device layer and stores and manages location data of the system. The motion manager provides user with stream data of field status and stores the data to the information storage. The information storage saves the information of the physical devices to database. Figure 2 describes the database schema.

The web server provides user with system information collecting from physical devices via internet.

3.1 Soil and Environment Monitoring Service

The soil and environment monitoring service shows data gathering from soil and environment sensors to user. Let's explain detail service procedure. First, this service transmits environment's raw data to the sensor manager. The raw data are temperature, humidity, Electronic Conductivity (EC) ratio, and illumination of field. Then, the sensor manager changes raw data into environment information and stores the information to the information storage. The web server sends the environment information stored in database to the user GUI. Figure 3 shows detail procedure of the soil and environment monitoring service.

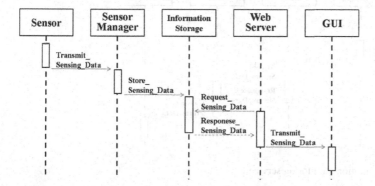

Fig. 3 Procedure of soil and environment monitoring service

3.2 Location Monitoring Service

This service monitors the UFSS's location in field. First, the GPS Module sends location data to the location manager. Then the sensor manager stores the data to the information storage. And the web server provides user with the location of the system by using GUI. You can see the procedure of this service in Figure 4.

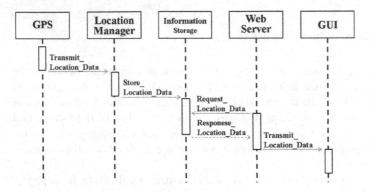

Fig. 4 Procedure of location monitoring service

3.3 Motion Monitoring Service

This service provides user with motion data through CCTV camera. First, the CCTV camera sends stream data to the motion manager. The motion manager stores the stream data to the information storage. User requests the stream data to the web server. Then the web server shows the stream data to the user via internet. Figure 5 describes the procedure of the motion monitoring service.

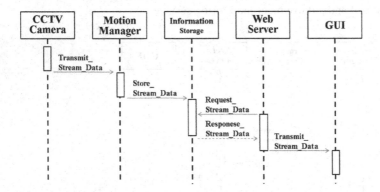

Fig. 5 Procedure of motion monitoring service

4 Implementation of the System

In this chapter, we develop the UFSS by implementing the system's components. Figure 6 is the system model. The UFSS is autonomous system. The system has solar cell and a storage battery. The system stores power in the daytime and uses in the nighttime.

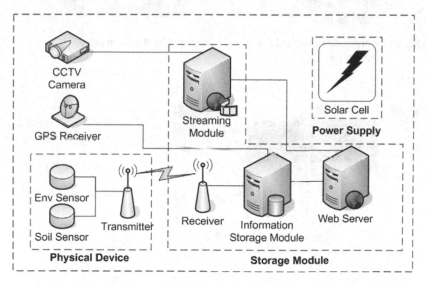

Fig. 6 UFSS System model

4.1 System Components

This system consists of physical devices and software modules. We explain the software module in Chapter 3. The physical devices divide into three parts. First, our system has sensing and information gathering devices. You can see the above devices in Figure 7.

And, you can see the sensors' data receiver, database, and the web server in Figure 8 and a solar battery in Figure 9.

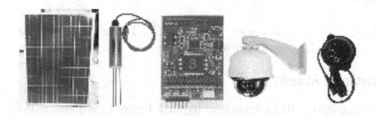

Fig. 7 Solar cell, soil sensor, network node, CCTV camera, and GPS module

Fig. 8 Sensor receiver, database, and the web server

Fig. 9 Solar battery

Fig. 10 Prototype of the UFSS

Now, we integrate above components into a system. Figure 10 shows the UFSS's physical devices including the software modules.

4.2 Implementation Results

We confirm the executing results of the system by GUI. Figure 11 is showing the UFSS's GUI. The (a) in the figure is showing the real-time motion from the CCTV camera in the system. The (b) is location of the system. We use the GPS

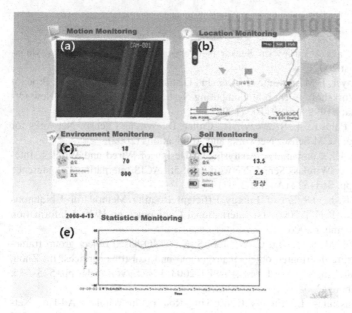

Fig. 11 UFSS's GUI

data to map the location. On the (c) is showing the sensing value from the soil sensor, and the (d) is the sensing value from the environment sensors. The average temperature is shown in (e).

5 Conclusions

This paper proposed the Ubiquitous Field Server System (UFSS) that could monitor environments of agricultural field. Also, for verifying the executability of our
system, we made a UFSS prototype and implemented system's components. Then
we showed the executing results of the system. From this result, we confirmed that
our system could monitor the field conditions by using various sensors and facilities. The problem executing the system is a power supply in nighttime. But, if you
use the battery storage with good performance, this problem will be solved.

In the future, we will make low-power and small-size system in order to use
commercial market and apply into the reference point control in GIS field.

Acknowledgements. This research was supported by the MKE (Ministry of Knowledge
Economy), Korea, under the ITRC(Information Technology Research Center) support
program supervised by the IITA (Institute of Information Technology Advancement) (IITA-
2009-(C1090-0902-0047)) and This work is financially supported by the Ministry of
Education, Science and Technology(MEST), the Ministry Knowledge Economy(MKE) and
the Ministry of Labor(MOLAB) through the Hub University for Industrial
Collaboration(HUNIC).

References

[1] Akyildiz, I.F., et al.: A survey on Sensor Networks. IEEE Communications Magazine 40(8) (August 2002)
[2] Perkins, C.E., Royer, E.M.: Ad-hoc On-Demand Distance Vector Routing. In: Proc. of the 2nd IEEE Workshop on Mobile Computing Systems and Applications, New Orleans, LA, February 1999, pp. 90–100 (1999)
[3] Mainwaring, A., Polastre, J., Szewczyk, R., Culler, D.: Wireless Sensor Networks for Habit Monitoring. ACM Sensor Networks and Applications (2002)
[4] Lee, M.-h., Yoe, H.: Comparative Analysis and Design of Wired and Wireless Integrated Networks for Wireless Sensor Networks. In: 5th ACIS International Conference on SERA 2007, pp. 518–522 (August 2007)
[5] Yoe, H., Eom, K.-b.: Design of Energy Efficient Routing Method for Ubiquitous Green Houses. In: ICHIT 2006, 1st International Conference on Hybrid Information Technology (November 2006)
[6] Shin, C.-S., Kang, M.-S., Jeong, C.-W., Joo, S.-C.: TMO-based object group framework for supporting distributed object management and real-time services. In: Zhou, X., Xu, M., Jähnichen, S., Cao, J. (eds.) APPT 2003. LNCS, vol. 2834, pp. 525–535. Springer, Heidelberg (2003)
[7] Chang, J.-H., Tassiulas, L.: Energy Conserving Routing in Wireless Ad-Hoc Networks. In: Proc. INFOCOM 2000, vol. 11(1), pp. 22–31 (March 2003); IEEE/ACM Trans. Networking 11(1), 2–16 (February 2003)

A Study on the Design of Ubiquitous Sensor Networks Based on Wireless Mesh Networks for Ubiquitous-Greenhouse

Jeong-Hwan Hwang, Hyun-Joong Kang, Hyuk-Jin Im, Hyun Yoe, and Chang-Sun Shin

Summary. Ubiquitous Sensor Networks (USNs) based on IEEE 802.15.4 is implemented in diversified fields such as environmental monitoring, disaster management, logistics management and home network, etc. By implementing such Ubiquitous Sensor Networks technology to Greenhouses it is possible to develop Ubiquitous-Greenhouse that allows environmental monitoring and actuator control. But, it is inadequate to operate large-scale network configured by expanding the network with limited energy resources of a sensor node. Therefore in this study, we virtually implemented Mesh coordinators between the sensor devices and data transmission paths toward a Gateway to solve above mentioned problem. In addition, to verify performance of the proposed USN based on Wireless Mesh Networks, we also conducted simulation on general USN based on IEEE 802.15.4 under identical environments then comparatively analyzed from acquired results of both simulations. Result has shown the performance of the proposed USN based on Wireless Mesh Networks (WMNs) having better performance, in terms of data transmission efficiency and energy consumption than USN based on IEEE 802.15.4.

Keywords: Wireless Sensor Networks, Ubiquitous Sensor Networks, Greenhouse, Wireless Mesh Networks.

1 Introduction

Ubiquitous Sensor Networks is the most essential core technology in Ubiquitous era for communicating without limitation in time, space and target by identifying objects, detecting physical phenomenon, etc. [1] In addition, Ubiquitous Sensor

Jeong-Hwan Hwang, Hyun-Joong Kang, Hyuk-Jin Im, Hyun Yoe,
and Chang-Sun Shin
School of Information and Communications Engineering, Sunchon National Univ.,
315 Maegok-dong, Sunchon-Si, Jeonnam 540-742, Korea
e-mail: {jhwang, hjkang, polyhj}@mail.sunchon.ac.kr,
{yhyun, csshin}@sunchon.ac.kr

Hyun Yoe
Corresponding author.

R. Lee, N. Ishii (Eds.): Software Engineering, Artificial Intelligence, SCI 209, pp. 255–262.
springerlink.com   © Springer-Verlag Berlin Heidelberg 2009

Networks provides computing and network functions to all forms of objects such as devices, machines, etc., By providing optimal functions such as automatic sensing of environmental status, etc., Ubiquitous Sensor Network is able to uplift comfort and safety to an optimal level for human life [2].

A sensor node is a component of USNs and operating with limited resources like a battery. It also has functions such as sensing, computing, wireless communication, etc., and freely form a network among nodes to acquire, process and control environmental data [3][4][5].

USNs are usually spread around in remote sites or hazardous area [6][7], each node has characteristics of free displacement, ability to form a network, high density, frequent network structure changes, limited energy and computing resources, point-to-point communication, etc. [3][4], thus they are inadequate for structuring a large-scale network.

In this study we proposed USN based on Wireless Mesh Networks implemented with IEEE 802.15.4 Standard for achieving enhanced data throughput and improved service quality under USN for conducting environmental monitoring.

In this study, relay node which is functioning as the most important bridge between a sensor device that collects data, to a Gateway which also called the final node, has been replaced with Mesh coordinators for building a Wireless Mesh Network then conducted simulation to verify its performance.

This paper is organized as follows. In Chapter 2, the proposed USN based on Wireless Mesh Network is explained. In Chapter 3, performance of USN based on IEEE 802.15.4 and the proposed USN based on Wireless Mesh Networks were comparatively analyzed by conducting simulations. Finally in Chapter 4, conclusion of the study will be stated.

2 Design of USN Based on Wireless Mesh Networks

2.1 The Issues of USN Based on IEEE 802.15.4

Currently, USN based on IEEE 802.15.4 is widely utilized for environmental monitoring. Under an environment without communication infrastructure, sensor devices forward data through other sensor nodes to perform long distance transmission via multi-hop in transmitting collected data to a distant sink node or a Gateway. In the case of long distance transmission, changes in network formation under Ubiquitous Sensor Network occur frequently thus possibility of packet loss becomes much bigger. Size of data under USN is small and has lesser frequency of data generation. By considering these factors, packet loss will become a serious problem in information collection.

In addition, sensor devices transmit data through broadcast communication thus nearby sensors that pass on such data tend to consume more power and have larger overhead. Resultantly, data throughput will be low and QoS(Quality of Service) can not be guaranteed.

2.2 Design of USN Based on Wireless Mesh Networks

In this study, we placed Mesh coordinators between the sensor device nodes and a Gateway which acts as the destination node of USN based on IEEE 802.15.4 to resolve mentioned issues.

Wireless Mesh Networks configured by utilizing Mesh Coordinators show high reliability in data transmission for the network utilizes multi-path method composed of Mesh links. Furthermore, the load put on the sensor devices during the data communication can be concentrated to Mesh coordinators instead. Moreover, sensor device, including environmental sensors, conducts tasks with limited energy resources than Mesh coordinators. By conducting data transmission with utilization of Mesh coordinators of the configured Wireless Mesh Network, energy consumed by sensor devices can be reduced greatly and overall lifetime of the network can be extended.

Fig. 1 shows the structure of energy efficient Ubiquitous Sensor Network with the use of the proposed Wireless Mesh Network.

When the simulation initiates, each Mesh coordinators configures Wireless Mesh Network with the Gateway as the core. When Mesh coordinators complete Wireless Mesh Network configuration, they search nearby sensor devices and make connections. By conducting point-to-point communication, they receive environmental data from sensor devices. Mesh coordinators transmit data received from sensor devices to the Gateway through multi-path.

Ubiquitous Sensor Networks configuration utilizing Wireless Mesh Networks can extend transmission distance while reducing energy consumption of sensor devices.

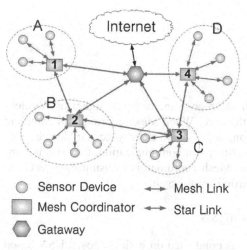

Fig. 1 Sensor Networks structure for Ubiquitous Greenhouse by using Wireless Mesh Networks

3 Simulation

In this study, we conducted simulation utilizing NS-2 on the proposed USN based on Wireless Mesh Networks and USN based on IEEE 802.15.4 under identical environments [8].

In addition, we acquired throughput of generated packet, throughput of dropped packet and residual energy of Mesh coordinators that acts as relay nodes in Wireless Mesh Networks and relay nodes of USN based on IEEE 802.15.4 during the procedure of transmitting data from sensor device to final Gateway.

3.1 Simulation Environment

In the simulation, sensor devices and Mesh coordinators for conducting environmental monitoring of Greenhouses were spread around in 3000 × 3000(m) area. By considering the fact that data collected by environmental monitoring may show large variation depending on the location within the Greenhouse [9], we placed 2 sensor devices within the Greenhouse to detect environmental changes.

Fig. 2 shows topology configuration for conducting simulations. For the simulation topology, real Greenhouses example in Suncheon Bay, Suncheon-si Jeollanamdo were referenced.

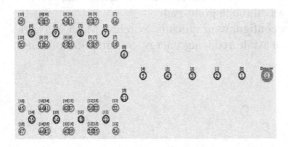

Fig. 2 Topology Configuration

When the simulation initiates, each Mesh coordinator try to organizes Wireless Mesh Network with the Gateway as the core. When Mesh coordinators complete Wireless Mesh Network configuration, they search nearby sensor devices and make connections. By conducting point-to-point communication, they receive environmental data from sensor devices. Mesh coordinators transmit data received from sensor devices to the Gateway through multi-path.

3.2 Simulation Results and Analysis

Here, we analyzed results of simulations conducted on both proposed USN based on Wireless Mesh Networks and USN based on IEEE 802.15.4.

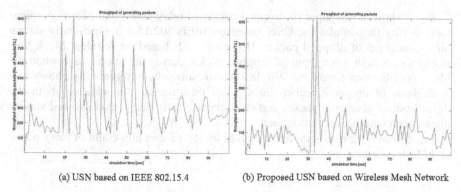

(a) USN based on IEEE 802.15.4 (b) Proposed USN based on Wireless Mesh Network

Fig. 3 Throughput of generated packet

Fig. 3 shows throughput of generated packet under respective networks.

Time taken for sensor devices to complete connection with relay nodes under USN based on IEEE 802.15.4 was 16 seconds. Under USN based on Wireless Mesh Networks, Mesh coordinators, which replace relay nodes, form a Mesh network first then initiate connection with sensor devices. The time taken to complete connection was 30 seconds.

In both network, the number of the generated data packets is not large at the initial stage when the sensor devices complete connection with relay nodes or Mesh coordinators.

In addition, USN based on IEEE 802.15.4 had shown shorter duration to complete connection than USN based on Wireless Mesh Networks.

During the simulation, USN based on IEEE 802.15.4 shows high packet throughput and large variation in generated packet. However, the study has confirmed USN based on Wireless Mesh Networks showing lesser Packet throughput and generated packet variation than USN based on IEEE 802.15.4. This is because under USN based on IEEE 802.15.4, relay nodes and sensor devices utilizes Broadcast Communication thus the number of the generated data packets increases.

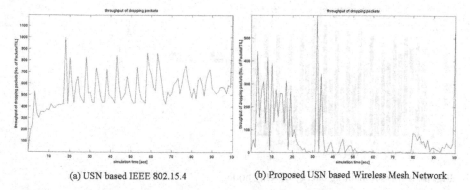

(a) USN based IEEE 802.15.4 (b) Proposed USN based Wireless Mesh Network

Fig. 4 Throughput of dropped packet

Fig. 4 shows throughput of dropped packet of relay nodes and Mesh coordinators. During the simulation, USN based on IEEE 802.15.4 has generally shown high throughput of dropped packet. However, USN based on Wireless Mesh Networks shows high throughput of dropped packet during the initial stage when the Mesh coordinators forms the Wireless Mesh Networks but generally shows low throughput of dropped packet during data transmission. In addition, both the throughput of generated packet and throughput of dropped packet showed similar phenomena with packet throughput.

Fig. 5 shows transmission delay caused by increase in hop count of relay nodes and Mesh coordinators.

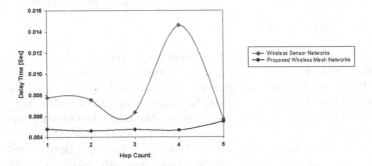

Fig. 5 Delay due to increased hop count

USN based on IEEE 802.15.4 tends to have relatively high transmission delay and transmission delay variation tends to show large change when the hop count increases. On the other hand, USN based on Wireless Mesh Networks tends to have relatively low transmission delay and transmission delay variation doesn't affect much even when hop count increases.

Fig. 6 shows average remaining energy of relay nodes in each network and average remaining energy of Mesh coordinators under WMN.

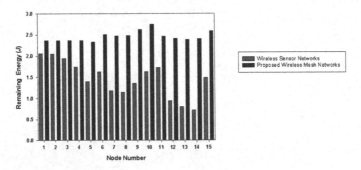

Fig. 6 Average remaining energy of relay nodes

USN based on IEEE 802.15.4 is used with broadcast communication therefore relay nodes consume more energy as data throughput and undesired activities increases. On the other hand, USN based on Wireless Mesh Networks is used with point-to-point communication thus relay nodes of USN based on Wireless Mesh Networks consumes less energy than relay nodes of USN based on IEEE 802.15.4.

Fig. 7 shows remaining energy of sensor devices under respective networks. Sensor devices of Under USN based on IEEE 802.15.4 consumed more energy than sensor devices of USN based on Wireless Mesh Networks.

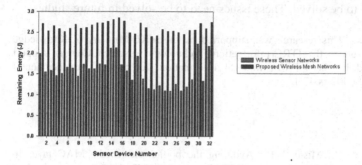

Fig. 7 Average remaining energy of sensor device

4 Conclusion

In this study, we proposed the USN based on Wireless Mesh Networks. To transmit data collected from sensor devices to long-distanced a sink node or a Gateway under USN based on IEEE 802.15.4, broadcast communication is used. This broadcast communication increases overhead of the network and increases power consumption of sensor devices. In addition, formation of USN changes frequently thus during the long distance data transmission, packet loss may occur. Consequently, data throughput will be greatly reduced and QoS can not be guaranteed.

To resolve mentioned problems, we virtually built Mesh coordinators between the sensor devices and data transmission paths toward a Gateway. By doing so, sensor devices can transmit data to Gateway by conducting point-to-point communication with surrounding Mesh coordinators without using broadcast communication. Here, Mesh coordinators form a Wireless Mesh Network among them and transmits data collected from sensor devices to Gateway through multi-path.

In this study, we conducted simulations on the proposed USN based on Wireless Mesh Networks and USN based on IEEE 802.15.4 under identical environments to comparatively analyze their performances in determining superiority in performance of the proposed USN based on Wireless Mesh Networks.

Results of simulation have shown the proposed USN based on Wireless Mesh Network having less average transmission delay and transmission delay due to increased hop count than USN based on IEEE 802.15.4. In addition, the proposed USN based on Wireless Mesh Network has shown lesser generated packet variation and higher packet transfer rate than USN based on IEEE 802.15.4.

Also average remaining energy of sensor devices and relay nodes have shown relatively higher and the study was able to determine increase in overall energy efficiency.

Consequently, the USN based on Wireless Mesh Networks proposed in this study has shown better results in terms of data transmission efficiency and energy efficiency than USN based on IEEE 802.15.4.

In order to realize the proposed USN based on Wireless Mesh Networks, such as extending the transmission distance between sensor devices and Mesh coordinators and reducing transmission delay caused by increase in hop count of Mesh coordinators have to be solved. These issues need to be solved in future studies.

Acknowledgements. "This research was supported by the MKE(Ministry of Knowledge Economy), Korea, under the ITRC(Information Technology Research Center) support program supervised by the IITA(Institute of Information Technology Advancement)" (IITA-2009-(C1090-0902-0047)).

References

[1] Misic, J., Shafi, J., Misic, V.B.: Avoiding the bottlenecks in the MAC layer in 802.15.4 low rate WPAN. In: Proc. of ICPADS, pp. 363–367 (2005)

[2] Kim, J.E., Kim, S.H., Jeong, W.C., Kim, N.S.: Technical Trend of USN sensor node. ETRI Telecommunication review 22(3), 90–103 (2007)

[3] Ye, W., Heidemann, J., Estrin, D.: An energy-efficient MAC protocol for wireless sensor networks. In: proceeding of the IEEE infocom, pp. 1567–1576 (2002)

[4] Kubisch, M., Karl, H., Wolisz, A., Zhong, S.C., Rabaey, J.: Distributed algorithms for transmission power control in wireless sensor networks. Proceeding of IEEE Wireless Communications and Networking 1, 558–563 (2003)

[5] Akyildiz, I.F., Su, W., Sankarasubramaniam, y., Cayirci, E.: A survey on sensor networks. IEEE Communications Magazine 40(8), 102–114 (2002)

[6] Shah, R.C., Rabaey, J.M.: Energy aware routing for low energy ad hoc sensor networks. In: IEEE Wireless Communications and Networking Conference, pp. 350–355 (March 2002)

[7] El-Hoiydi, A.: Spatial TDMA and CSMA with preamble sampling for low power ad hoc wireless sensor networks. In: proceeding of the Seventh International Symposium on Computers and Communications (ISCC 2002), pp. 685–692 (July 2002)

[8] Fall, K., Varadhan, K.: The ns Manual (formerly ns Notes and Documentation). The VINT Project, A Collaboration between researches at UC Berkeley, LBL, USC/ISI and Xerox PARC

[9] Soon-Joo, K., Su-Yeun, N.: A Study on the Thermal Environment in the Multipurpose Greenhouse in Winter. Korean Solar Energy Society 27(3), 15–21 (2007)

[10] Perkins, C., Royer, E., Das, S.: Ad Hoc On-Demand Distance Vector (AODV) Routing. IETF Internet Draft (March 2000),
http://www.ietf.org/Internet-drafts/draft-ietf-manet-aodv-05.txt

Switching Software: A Hierarchical Design Approach

Byeongdo Kang and Roger Y. Lee

Abstract. In this paper we present an architecture style for switching software. In general, switching software includes application software, operating systems, and database management systems. Application software consists of call handling, operation, administration, and maintenance functions. To improve maintainability, we suggest a hierarchical structure based on the characteristics of switching software.

1 Introduction

The switching software that controls a switching system performs a complex job, not only because of the many different functions it must perform but because of the large number of functions that must be performed concurrently and with real-time requirements. Some requirements for the switching software are listed below [1]:

- Modularity: the switching software has hierarchical structures. The modular design allows increases in both network and processor capacity in reasonable and cost-effective increments.
- Reliability: the switching system is highly reliable. Automatic fault detection, fault location, and reconfiguration capabilities ensure that faults can be identified, isolated, and repaired timely, thereby providing better service at lower maintenance cost.
- Finite state machines: a finite state machine is a basic module that performs a switching function. Switching systems may be defined by means of one or more finite state machines with message protocols.

Byeongdo Kang
Department of Computer and Information Technology,
Daegu University,
Naeri 15, Jinryang, Gyeongsan, Gyeongbuk, Republic of Korea
e-mail: bdkang@daegu.ac.kr

Roger Y. Lee
Software Engineering and Information Technology Institute,
Central Michigan University,
Mt Pleasant, MI 48859, U.S.A.
e-mail: leelry@cmich.edu

R. Lee, N. Ishii (Eds.): Software Engineering, Artificial Intelligence, SCI 209, pp. 263–270.
springerlink.com   © Springer-Verlag Berlin Heidelberg 2009

- Half-call: most of the call processing software is designed with Half-call concepts. Every call is divided into two parties; an originating and a termination party.
- Maintainability: the life time of a switching system is very long. Its life time is twenty years or more. This life time imposes the need for effective management of stored program controlled software to cover the changes that will be necessitated by service requirements during such a long life time. It is, therefore, important to design software architecture with high maintainability.

In section 2, we show the general structure of switching software. Section 3 introduces the decomposition method for telephony software. In section 4, we conclude with a summary.

2 General Structure of Switching Software

The switching software system that runs in telecommunication networks may evolve continuously by three reasons; the addition of new functions, the correction of errors, or the adaptation of operating environments. It is, therefore, important to design software architecture with high maintainability.

To improve its maintainability, the switching software can be structured into a layered style shown in figure 1. General switching software consists of four common functions: operating systems, DBMS, man-machine interfaces, and telephony software such as administration, call handling, and maintenance.

Operating systems includes functions such as process management, IPC management, time management, memory management, exception handlers, I/O management, and file management. DBMS keeps data that application software generates or uses. Man-machine interfaces are provided for operators who

Fig. 1 General Structure of Switching Software

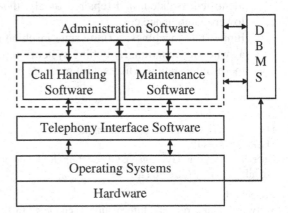

communicate with switching machines. Telephony software consists of call handling, administration, and maintenance software. Call handling software includes the functions such as call and channel processing and device control. Administration and maintenance software include accounting, measurement and statistics, network management, and data processing functions.

3 Functional Decomposition Method for Telephony Software

Real-time Structured Analysis [2] is used as the first step of our decomposition modeling. The functional decomposition method is based on the functional decomposition and concurrent task structuring.

3.1 Decomposition Strategies

We use Real-Time Structured Analysis as the first step of our specification method, and extend the concept of concurrent tasks. Our specification method is based on the following strategies;

- Functional decomposition: The software system is decomposed into functions and interfaces between them are defined in the form of data flows or control flows. Functions are mapped onto processors, tasks, and modules.
- Concurrent tasks: Concurrent tasking is considered a key aspect in concurrent and real-time design. We provide a guideline for structuring switching software into concurrent tasks.

3.2 Decomposition Concepts

The software system is decomposed into functions and the interfaces between them are defined in the form of data flows or control flows. Functions are represented by transformations. Transformations may be data or control transformations. Functions are mapped to modules to be implemented.

Finite state machines are used to define the behavioural characteristics of the system. State transition diagrams are an effective tool to represent finite state machines. A control transformation represents the execution of a state transition diagram.

A task represents the execution of a sequential component in a concurrent system. Each task deals with one sequential thread of execution. Thus, no concurrency is allowed within a task. The tasks often execute asynchronously, and are independent of each other.

Abstraction is used as a criterion for encapsulating data stores. Where an abstraction is accessed by more than one task, the access procedures must synchronize the access of data.

Fig. 2 A Context Diagram
Example

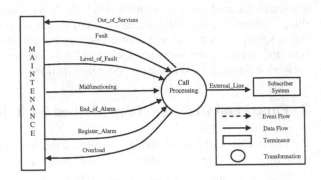

3.3 Decomposition Steps

The steps in our specification method are as follows:

(1) Develop a system essential model

The system context diagram and state transition diagrams are developed. The context diagram is decomposed into hierarchically data flow and control flow diagrams. The relationship between the state transition diagrams and the control and data transformations is developed. A control transformation is defined by means of a state transition diagram. Only event flows are inputs to or outputs from a control transformation. Input event flows to a control transformation may be trigger event flows from the external environment or from data transformations. Output event flow from a control transformation may be trigger event flows to the external environment. A data transformation may be activated by a trigger or enable event flow from a control transformation, a discrete data flow or event flow from another data transformation or a timer event. Figure 2 shows an example of a simple context diagram. The context diagram is a description of the environment in which the system operates. This diagram has two pieces; a description of the boundary between the system and the environment, showing the interfaces between the two parts, and a description of the events that occur in the environment to which the system must respond.

(2) Allocate transformations to processors

The transformations in the essential model are allocated to the processors of the target system. If necessary, the data flow diagrams are redrawn for each processor. Each processor is responsible for some portion of the essential model.

(3) Allocate transformations to tasks

The transformations for each processor are allocated to concurrent tasks. Each task represents a sequential program. Some task structuring categories are shown in [3]; the I/O task structuring, the internal task structuring, the task cohesion, and the task priority criteria. I/O data transformations are mapped to asynchronous I/O tasks, periodic I/O tasks, or resource monitor tasks. Internal transformations are mapped to control, periodic, or asynchronous tasks and may be combined with

Fig. 3 A Hierarchical Decomposition Example

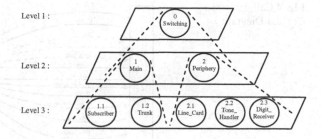

other transformations according to the sequential, temporal, or functional cohesion criteria. With the previous categories, we provide a guideline for structuring switching software into concurrent tasks. Using the Half-call characteristics of the switching software, we divide the switching software into an originating and a terminating party. Each party is decomposed into functions and interfaces between them are defined in the form of data flows or event flows. Functions are mapped onto processors and tasks. At this stage, a timing analysis may be performed.

The required response times are allocated to each task. Using the Structured Design method [4], each task is structured into modules. The function of each module and its interface to other modules are defined. Figure 3 shows the decomposition of a switching system into five tasks: subscriber, Trunk, Line_Card, Tone_Handler, and Digit_Receiver.

3.4 Specification Products

Because Real-Time Structured Analysis is used as the first step of our decomposition method, an environmental model and a behavioral model are developed. The environmental model describes the environment within which the system and outputs from the system. The behavioral model describes the responses of the system to the inputs it receives from the external environment.

(1) Structure Representation

The environmental model defines the boundary between the system to be developed and the external environment. It is described by means of the system context diagram, which shows the external entities that the system has to interface to, as well as the inputs to the system and outputs from the system. Inputs from and outputs to the external entities are represented by data flows. An event flow is a discrete signal and has no data value. It is used to indicate that some event has happened or to indicate a command. A terminator usually represents a source of data or a sink of data. In some cases, the system to be developed is actually a large subsystem that has to interface to other subsystems. In this case, the context diagram actually represents the boundary between the subsystem and the other subsystems and external entities it interfaces to. If a context diagram is used to represent a subsystem, then any external subsystem that this subsystem has to interface to should be shown as a terminator. A simple system context diagram for the Call Processing example is given in Figure 4.

Fig. 4 Call_Processing
Context Diagram

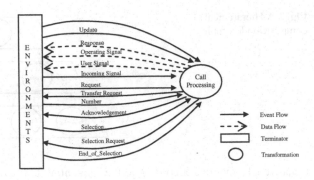

The behavioral model is represented by a hierarchical set of data flow or control flow diagrams. At non-leaf levels of the hierarchy, a data transformation is used to represent a subsystem, which may in turn be decomposed into lower-level subsystems. At the levels of the hierarchy, control transformations are used to represent control logics, while data transformations are used to represent non-control logics. The control aspects are represented by state transition diagrams. A state transition diagram consists of states and transitions. States are represented by rectangles while the labeled arrows represent transitions. The transition is associated with the input event that causes the state transition. Optionally, there may be a Boolean condition, while must be true when the event occurs for the state transition to take place. Figure 5 is a behavioral model represented by a hierarchical set of data flow diagrams. The control logics of the originating Half call of the Call_Processing example is represented by the state transition diagram given in Figure 6.

(2) Process Specifications
Leaf-level data transformations are defined by means of process specifications. A control transformation doesn't need a process specification because it is fully defined by means of state transition diagram. As one of process specifications, we use the process diagram notation given in SDL. An example of a process specification for a data transformation, Call_Transfer_Account in Accounting, is given in Figure 7.

(3) Data definition
A data dictionary is used to define all data and event flows. Entity-relationship diagrams [5] are used to show the relationships between the data stores of the system. They are used for identifying the data stores and for defining the contents of the stores. A definition example of the data given in Figure 4 is listed below.

- Signal_User_Incomming =
  Incomming_Signals + Request + Number + Selection + End_of_Selection,
- Signal_User_outgoing =
  User_Signal + Transfer_Request + Acknowledgement + Selection_Request,
- Signal_Operator_Incomming =
  Operator_Signals + Update + Info_Request,
- Signal_Operator_ Outgoing =
  Responses.

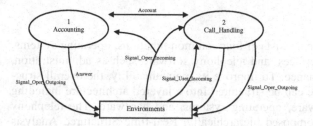

Fig. 5 Call_Processing System:Decomposition into Subsystems

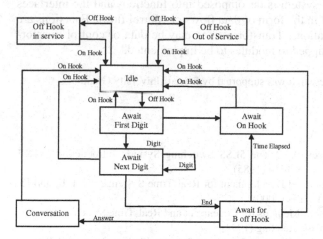

Fig. 6 State Transition Diagram

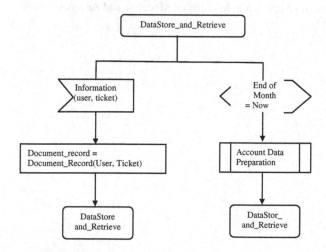

Fig. 7 Process Diagram

4 Conclusions

General switching software consists of four common functions: operating systems, DBMS, man-machine interfaces, and telephony software such as administration, call handling, and maintenance. To improve its maintainability, the overall structure for switching software can be structured into a layered architecture including telephony interfaces software, operating systems, and hardware. The telephony software also can be decomposed hierarchically. Real-time Structured Analysis was used as the first step of our decomposition modeling. The functional decomposition method is based on the functional decomposition and concurrent task structuring. The software system is decomposed into functions and the interfaces between them are defined in the form of data flows or control flows. Functions are represented by transformations. Transformations may be data or control transformations. Functions are mapped to modules to be implemented.

Acknowledgements. This research was supported by Daegu University Grant.

References

1. Martersteck, K.E., Spencer, A.E.: The 5ESS Switching System: Introduction. AT&T Technical Journal 64(6), 1305–1314 (1985)
2. Ward, P., Meller, S.: Structured Development for Real-Time Systems, vol. I, II, and III. Prentice-Hall, Englewood Cliffs (2000)
3. Gomaa, H.: Software Design Methods for Concurrent and Real-Time Systems. Addison-Wesley Publishing Company, Reading (1993)
4. Page-Jones, M.: The Practical Guide to Structured Systems Design. Prentice-Hall International Inc., Englewood Cliffs (1988)
5. Chen, P.P.: Entity-Relationship Approach to Information Modeling and Analysis. North-Holland, Amsterdam (1983)

Towards Data-Driven Hybrid Composition of Data Mining Multi-agent Systems

Roman Neruda

Abstract. The approach of solving data mining tasks by a collection of cooperating agents can profit from modularity, interchangeable components, distributed execution, and autonomous operation. The problem of automatic configuration of agent collections is studied in this paper. A solution combining logical resolution system and evolutionary algorithm is proposed and demonstrated on a simple example.

1 Introduction

The approach of solving data mining tasks by a collection of cooperating agents can profit from modularity, interchangeable components, distributed execution, and autonomous operation. Autonomous agents are small self-contained programs that can solve simple problems in a well-defined domain [16]. In order to solve complex data mining tasks, agents have to collaborate, forming multi-agent systems (MAS). A key issue in MAS research is how to generate MAS configurations that solve a given problem [7]. In most systems, a human user is required to set up the system configuration. Developing algorithms for automatic configuration of Multi-Agent Systems is a major challenge for AI research.

We have developed a platform for creating Multi-Agent Systems [12], [15]. Its main areas of application are computational intelligence methods (genetic algorithms, neural networks, fuzzy controllers) on single machines and clusters of workstations. Hybrid models, including combinations of artificial intelligence methods such as neural networks, genetic algorithms and fuzzy logic controllers, seem to be a promising and extensively studied research area [3]. Our distributed multi-agent system — provides a support for an easy creation and execution of such hybrid AI models utilizing the Java/JADE environment.

Roman Neruda
Institute of Computer Science, Academy of Sciences of the Czech Republic,
Pod vodárenskou věží 2, 18207 Prague 8, Czech Republic,
e-mail: roman@cs.cas.cz

R. Lee, N. Ishii (Eds.): Software Engineering, Artificial Intelligence, SCI 209, pp. 271–281.
springerlink.com   © Springer-Verlag Berlin Heidelberg 2009

The above mentioned applications require a number of cooperating agents to fulfill a given task. So far, MAS are created and configured manually. In this paper, we introduce two approaches for creation and possible configuration of MAS. One of them is based on formal descriptions and provides a logical reasoning component for the system.

The second approach to MAS generation employs evolutionary algorithm (EA) which is tailored to work on special structures—directed acyclic graphs—denoting MAS schemata. The advantage of EA is that it requires very little additional information apart from a measure of MAS performance. Thus, the typical run of EA consists of thousands of simulations which build and assess the fitness values of various MAS. Since the properties of logical reasoning search and evolutionary search are dual, the ultimate goal of this work is to provide a solution combining these two approaches in a hybrid search algorithm. This paper presents the first steps towards such a goal.

2 Related Work

There is a lot of research in how to use formal logics to model ontologies. The goal of this research is to find logics that are both expressive enough to describe ontological concepts, and weak enough to allow efficient formal reasoning about ontologies.

The most natural approach to formalize ontologies is the use of First Order Predicate Logics (FOL). This approach is used by well known ontology description languages like Ontolingua [8] and KIF [10]. The disadvantage of FOL-based languages is the expressive power of FOL. FOL is undecidable [6], and there are no efficient reasoning procedures. Nowadays, the de facto standard for ontology description language for formal reasoning is the family of description logics. Description logics are equivalent to subsets of first order logic restricted to predicates of arity one and two [5]. They are known to be equivalent to modal logics [1]. For the purpose of describing multi-agent systems, description logics are sometimes too weak. In these cases, we want to have a more expressive formalism. We decided to use Prolog-style logic programs for this. In the following we describe how both approaches can be combined together.

Description logics and Horn rules are orthogonal subsets of first order logic [5]. During the last years, a number of approaches to combine these two logical formalisms in one reasoning engine have been proposed. Most of these approaches use tableaux-style reasoners for description logics and combine them with Prolog-style Horn rules. In [11], Hustadt and Schmidt examined the relationship between resolution and tableaux proof systems for description logics. Baumgartner, Furbach and Thomas propose a combination of tableaux based reasoning and resolution on Horn logic [2]. Vellion [17] examines the relative complexity of SL-resolution and analytic tableau. The limits of combining description logics with horn rules are examined by Levy and Rousset [13]. Borgida [4] has shown that Description Logics and Horn rules are orthogonal subsets of first oder logic.

3 Computational Agents

An *agent* is an entity that has some form of perception of its environment, can act, and can communicate with other agents. It has specific skills and tries to achieve goals. A *Multi-Agent System (MAS)* is an assemble of interacting agents in a common environment [9]. In order to use automatic reasoning on a MAS, the MAS must be described in formal logics. For the computational system, we define a formal description for the static characteristics of the agents, and their communication channels. We do not model dynamic aspects of the system yet.

Agents communicate via messages and triggers. Messages are XML-encoded FIPA standard messages. Triggers are patterns with an associated behavior. When an agent receives a message matching the pattern of one of its triggers, the associated behavior is executed. In order to identify the receiver of a message, the sending agent needs the message itself and an id of the receiving agent. A conversation between two agents usually consists of a number of messages conforming to FIPA protocols. For example, when a neural network agent requests training data from a data source agent, it may send the following messages:

- Open the data source,
- Randomize the order of the data items,
- Set the cursor to the first item,
- Send next item.

These messages belong to a common category: Messages requesting input data from a data source. In order to abstract from the actual messages, we subsume all these messages under a *message type* when describing an agent in formal logics.

A *message type* identifies a category of messages that can be send to an agent in order to fulfill a specific task. We refer to message types by unique identifiers. The set of message types understood by an agent is called its *interface*. For outgoing messages, each link of an agent is associated with a message type. Via this link, only messages of the given type are sent. We call a link with its associated message type a *gate*.

Now it is easy to define if two agents can be connected: Agent A can be connected to agent B via gate G if the message type of G is in the list of interfaces of agent B. Note that one output gate sends messages of one type only, whereas one agent can receive different types of messages. This is a very natural concept: When an agent sends a message to some other agent via a gate, it assigns a specific role to the other agent, e.g. being a supplier of training data. On the receiving side, the receiving agent usually should understand a number of different types of messages, because it may have different roles for different agents. A connection is described by a triple consisting of a sending agent, the sending agent's gate, and a receiving agent.

Next we define *agents* and *agent classes*. Agents are created by generating instances of classes. An agent derives all its characteristics from its class definition.

Table 1 Concepts and roles used to describe MAS

Concepts	
mas(C)	C is a Multi-Agent System
class(C)	C is the name of an agent class
gate(C)	C is a gate
m_type(C)	C is a message type
Roles	
type(X,Y)	Class X is of type Y
has_gate(X,Y)	Class X has gate Y
gate_type(X,Y)	Gate X accepts messages of type Y
interface(X,Y)	Class X understands mess. of type Y
instance(X,Y)	Agent X is an instance of class Y
has_agent(X,Y)	Agent Y is part of MAS X

Additionally, an agent has a name to identify it. The static aspects of an agent class are described by the interface of the agent class (the messages understood by the agents of this class), the gates of the agent (the messages send by agents of this class), and the type(s) of the agent class. Types are nominal identifiers for characteristics of an agent. The types used to describe the characteristics of the agents should be ontological sound.

An *agent class* is defined by an interface, a set of message types, a set of gates, and a set of types. An *agent* is an instance of an agent class. It is defined by its name and its class.

4 Multi-agent Systems

A Multi-Agent System can be described by three elements: The set of agents in the MAS, the connections between these agents, and the characteristics of the MAS. The characteristics (constraints) of the MAS are the starting point of logical reasoning: In *MAS checking* the logical reasoner deduces if the MAS fulfills the constraints. In *MAS generation*, it creates a MAS that fulfills the constraints, starting with an empty MAS, or a manually constructed partial MAS. *Multi-Agent Systems (MAS)* consist of a set of agents, a set of connections between the agents, and the characteristics of the MAS.

Description logics know concepts (unary predicates) and roles (binary predicates). In order to describe agents and Multi-Agent Systems in description logics, the above notions are mapped onto description logic concepts and roles as shown in table 1. An example of the multilayer perceptron agent class with one gate and one interface follows:

```
class(multilayer_perceptron)
type(multilayer_perceptron, computational_agent)
has_gate(multilayer_perceptron, data_in)
gate_type(data_in, training_data)
interface(multilayer_perceptron, control_messages)
```

In the same way, instances of a particular agent classes are defined: *instance(multilayer_perceptron, mlp_instance)*. An agent is assigned to a MAS via role "has_agent". In the following example, we define "mlp_instance" as belonging to MAS "my_mas": *has_agent(my_mas, mlp_instance)*.

Since connections are relations between three elements, a sending agent, a sending agent's gate, and a receiving agent, we can not formulate this relationship in traditional description logics. It would be possible to circumvent the problem by splitting the triple into two relationships, but this would be counter-intuitive to our goal of defining MAS in an ontological sound way. Connections between agents are relationships of arity three: Two agents are combined via a gate. Therefore, we do not use description logics, but traditional logic programs in Prolog notation to define connections: *connection(mlp_instance, other_agent, gate)*

Constraints on MAS can be described in Description Logics, in Prolog clauses, or in a combination of both. As an example, the following concept description requires the MAS "mlp_MAS" to contain a decision tree agent: $mlp\_MAS \sqsupseteq mas \sqcap has\_agent.(\exists instance.decision\_tree)$

An essential requirement for a MAS is that agents are connected in a sane way: An agent should only connect to agents that understand its messages. According to connection definition above, a connection is possible if the message type of the sending agent's output gate matches a message type of the receiving agent s interface. With the logical concepts and descriptions given in this section, this constraint can be formulated as a Prolog style horn rule. If we are only interested in checking if a connection satisfies this property, the rule is very simple:

```
connection(S,R,G) ←
    instance(R, RC) ∧
    instance(S, SC) ∧
    interface(RC, MT)∧
    has_gate(SC, G) ∧
    gate_type(G, MT)
```

The first two lines of the rule body determine the classes *RC* and *SC* of the sending agent *S* and the receiving agent *R*. The third line instantiates *MT* with a message type understood by *RC*. The fourth line instantiates *G* with a gate of class *SC*. The last line assures that gate *G* matches message type *MT*.

The following paragraphs show an example for logical descriptions of MAS. It should be noted that these MAS types can be combined, i.e. it is possible to query for trusted, computational MAS. *Computational MAS:* A computational MAS can be defined as a MAS with a task manager, a computational agent and a data source agent which are interconnected.

comp_MAS(MAS) ←
    type(CAC, computational_agent)∧
    instance(CA, CAC)∧
    has_agent(MAS, CA)∧
    type(DSC, data_source)∧
    instance(DS, DSC)∧
    has_agent(MAS, DS)∧
    connection(CA, DS, G)∧
    type(TMC, task_manager)∧
    instance(TMC, TM)∧
    has_agent(MAS, TM)∧
    connection(TM, CA, GC)∧
    connection(TM, DS, GD)

Trusted MAS: We define that an MAS is trusted if all of its agents are instances of a "trusted" class. This examples uses the Prolog predicate findall. findall returns a list of all instances of a variable for which a predicate is true. In the definition of predicate all_trusted the usual Prolog syntax for recursive definitions is used.

trusted_MAS(MAS) ←
    findall(X, has_agent(MAS,X), A))∧
    all_trusted(A)
all_trusted([]) ← true
all_trusted([F|R]) ←
    instance(F,FC)∧
    type(FC, trusted) ∧
    all_trusted([R])

5 Evolutionary Search

The proposed evolutionary algorithm operates on schemes definitions in order to find a suitable scheme solving a specified problem. The genetic algorithm has three inputs: First, the number and the types of inputs and outputs of the scheme. Second, the *training set*, which is a set of prototypical inputs and the corresponding desired outputs, it is used to compute the fitness of a particular solution. And third, the list of types of building blocks available for being used in the scheme.

We supply three operators that would operate on graphs representing schemes: *random scheme creation*, *mutation* and *crossover*. The aim of the first one is to create a random scheme. This operator is used when creating the first (random) generation. The diversity of the schemes that are generated is the most important feature the generated schemes should have. The 'quality' of the scheme (that means whether the scheme computes the desired function or not) is insignificant at that moment, it is a task of other parts of the genetic algorithm to assure this. The algorithm for random scheme creation works incrementally. In each step one building block

Fig. 1 Crossover of two schemes. Parents are cut in a valid places and one offspring is created from the swapped parts

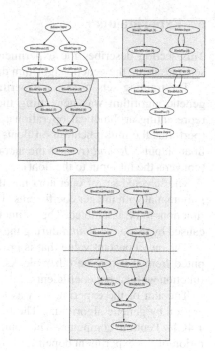

Fig. 2 Mutation on scheme. The destination of two corresponding links are switched

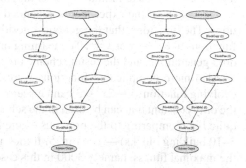

is added to the scheme being created. In the beginning, the most emphasis is put on the randomness. Later the building blocks are selected more in fashion so it would create the scheme with the desired number and types of gates (so the process converges to the desired type of function). The goal of the crossover operator is to create offsprings from two parents. The crossover operator proposed for scheme generation creates one offspring. The operator horizontally divides the mother and the father, takes the first part from father's scheme, and the second from mother's one. The crossover is illustrated in Fig. 1.

The mutation operator is very simple. It finds two links in the scheme (of the same type) and switches their destinations. The mutation operator is illustrated in Fig. 2.

6 Experiments

This section describes the experiments we have performed with generating the schemes using the genetic algorithm described above.

The training sets used for experiments represented various polynomials. The genetic algorithm was generating the schemes containing the following agents representing arithmetical operations: *Plus* (performs the addition on floats), *Mul* (performs the multiplication on floats), *Copy* (copies the only input (float) to two float outputs), *Round* (rounds the incoming float to the integer) and finally *Floatize* (converts the int input to the float).

The selected set of operators has the following features: it allows to build any polynomial with integer coefficients. The presence of the *Round* allows also another functions to be assembled. These functions are the 'polynomials with steps' that are caused by using the *Round* during the computation.

The only constant value that is provided is -1. All other integers must be computed from it using the other blocks. This makes it more difficult to achieve the function with higher coefficients.

The aim of the experiments was to verify the possibilities of the scheme generation by genetic algorithms. The below mentioned examples were computed on 1.4GHz Pentium computers. The computation is relatively time demanding. The duration of the experiment depended on many parameters. Generally, one generation took from seconds to minutes to be computed.

The results of the experiments depended on the complexity of the desired functions. The functions, that the genetic algorithm learned well and quite quickly were functions like $x^3 - x$ or $x^2 y^2$. The learning of these functions took from tens to hundred generations, and the result scheme precisely computed the desired function.

Also more complicated functions were successfully evolved. The progress of evolving function $x^3 - 2x^2 - 3$ can be seen in the Fig. 3 and 4. Having in mind, that the only constant that can be used in the scheme is -1, we can see, that the scheme is quite big (comparing to the previous example where there was only approximately 5–10 building blocks) — see Fig. 5. It took much more time/generations to achieve the maximal fitness, namely 3000 in this case.

Fig. 3 Function $x^3 - 2x^2 - 3$. The history of the maximal and average fitness

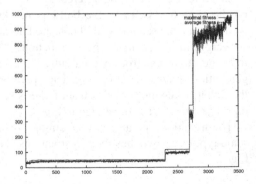

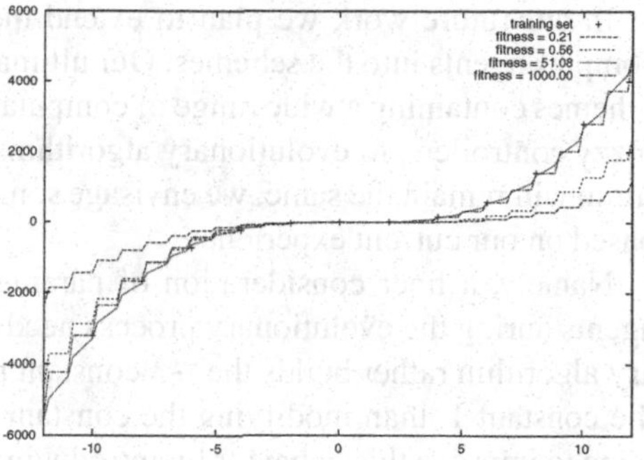

Fig. 4 Function $x^3 - 2x^2 -$
3. The best schemes from
generation 0, 5, 200 and
3000

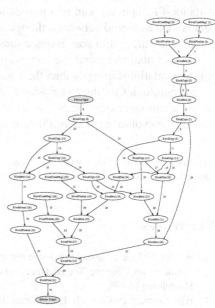

Fig. 5 Function $x^3 - 2x^2 -$
3. The scheme with fitness
1000 (out of 1000), taken
from 3000th generation

On the other hand, learning of some functions remained in the local maxima, which was for example the case of the function $x^2 + y^2 + x$.

7 Conclusions

We have presented a hybrid system that uses a combination of evolutionary algorithm and a resolution system to automatically create and evaluate multi-agent schemes. So far, the implementation has focused on relatively simple agents computing parts of arithmetical expressions. Nevertheless, the sketched experiments demonstrate the soundness of the approach. A similar problem is described and tackled in [18] by means of matchmaking in middle-agents where authors make use of ontological descriptions but utilize other search methods than EA.

In our future work we plan to extend the system in order to incorporate more complex agents into the schemes. Our ultimate goal is to be able to propose and test schemes containing a wide range of computational methods from neural networks to fuzzy controllers, to evolutionary algorithms. While the core of the proposed algorithm will remain the same, we envisage some modifications in the genetic operators based on our current experience.

Namely, a finer consideration of parameter values, or configurations, of basic agents during the evolutionary process needs to be addressed. So far, the evolutionary algorithm rather builds the -3 constant by combining three agents representing the constant 1, than modifying the constant agent to represent the -3 directly. We hope to improve this behavior by introducing another kind of genetic operator. This mutation-like operator can be more complicated in the case of real computational agents such as neural networks, though. Nevertheless, this approach can reduce the evolutionary algorithm search space substantially.

We also plan to extend the capabilities of the resolution system towards more complex relationship types than the ones described in this paper. In our work [14] we use ontologies for the description of agent capabilities, and have the CSP-solver reason about these ontologies. The next goal is to provide hybrid solution encompassing the evolutionary algorithm enhanced by ontological reasoning.

Acknowledgements. This work has been supported by the Ministry of Education of the Czech Republic under the Center of Applied Cybernetics project no. 1M684077004 (1M0567).

References

1. Baader, F.: Logic-based knowledge representation. In: Wooldrige, M.J., Veloso, M. (eds.) Artificial Intelligence Today, Recent Trends and Developments, pp. 13–41. Springer, Heidelberg (1999)
2. Baumgartner, P., Furbach, U., Thomas, B.: Model-based deduction for knowledge representation. In: Proceedings of the International Workshop on the Semantic Web, Hawaii, USA (2002)
3. Bonissone, P.: Soft computing: the convergence of emerging reasoning technologies. Soft computing 1, 6–18 (1997)
4. Borgida, A.: On the relationship between description logic and predicate logic. In: CIKM, pp. 219–225 (1994)
5. Borgida, A.: On the relative expressiveness of description logics and predicate logics. Artificial Intelligence 82(1–2), 353–367 (1996)
6. Davis, M. (ed.): The Undecidable—Basic Papers on Undecidable Propositions, Unsolvable Problems and Computable Functions. Raven Press (1965)
7. Doran, J.E., Franklin, S., Jennings, N.R., Norman, T.J.: On cooperation in multi-agent systems. The Knowledge Engineering Review 12(3), 309–314 (1997)
8. Farquhar, A., Fikes, R., Rice, J.: Tools for assembling modular ontologies in ontolingua. Technical report, Stanford Knowledge Systems Laboratory (1997)
9. Ferber, J.: Multi-Agent Systems: An Introduction to Distributed Artificial Intelligence. Addison-Wesley Longman, Harlow (1999)

10. Genesreth, M.R., Fikes, R.E.: Knowledge interchange format, version 2.2. Technical report, Computer Science Department, Stanford University (1992)
11. Hustadt, U., Schmidt, R.A.: On the relation of resolution and tableaux proof system for description logics. In: Thomas, D. (ed.) Proceedings of the 16th Internatoinal joint Conference on Artificial Intelligence IJCAI 1999, Stockholm, Sweden, vol. 1, pp. 110–115. Morgan Kaufmann, San Francisco (1999)
12. Krušina, P., Neruda, R., Petrova, Z.: More autonomous hybrid models in bang. In: International Conference on Computational Science (2), pp. 935–942. Springer, Heidelberg (2001)
13. Levy, A.Y., Rousset, M.C.: The limits of combining recursive horn rules with description logics. In: Proceedings of the Thirteenth National Conference on Artificial Intelligence, Portland, OR (1996)
14. Neruda, R., Beuster, G.: Towards dynamic generation of computational agents by means of logical descriptions. In: MASUPC 2007 – International Workshop on Multi-Agent Systems Challenges for Ubiquitous and Pervasive Computing, pp. 17–28 (2007)
15. Neruda, R., Krušina, P., Kudova, P., Beuster, G.: Bang 3: A computational multi-agent system. In: Proceedings of the 2004 WI-IAT 2004 Conference. IEEE Computer Society Press, Los Alamitos (2004)
16. Nwana, H.S.: Software agents: An overview. Knowledge Engineering Review 11(2), 205–244 (1995)
17. Vellino, A.: The relative complexity of sl-resolution and analytical tableau. Studia Logica 52(2), 323–337 (1993)
18. Zhang, Z., Zhang, C.: Agent-Based Hybrid Intelligent Systems. Springer, Heidelberg (2004)

Index

Author Index